70121185

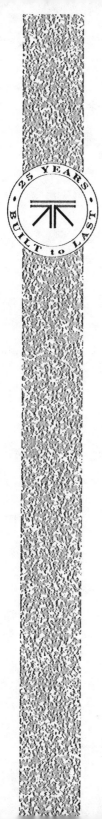

Also by William F. Allman

Apprentices of Wonder

Newton at the Bat

by
William F. Allman

A Touchstone Book
Published by Simon & Schuster

New York   London   Toronto
Sydney   Tokyo   Singapore

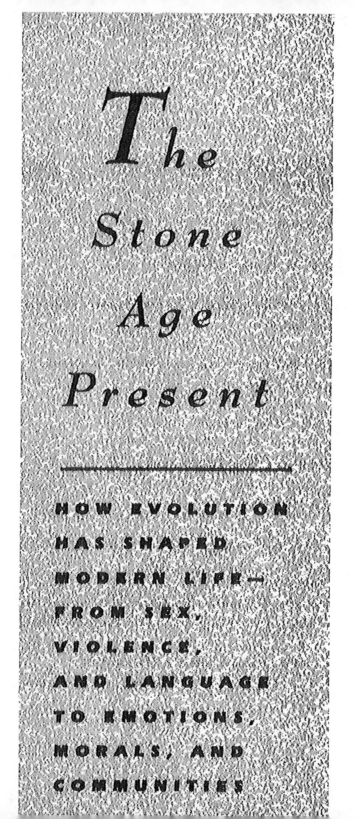

The Stone Age Present

HOW EVOLUTION
HAS SHAPED
MODERN LIFE—
FROM SEX,
VIOLENCE,
AND LANGUAGE
TO EMOTIONS,
MORALS, AND
COMMUNITIES

TOUCHSTONE
Rockefeller Center
1230 Avenue of the Americas
New York, New York 10020

First Touchstone Edition 1995

TOUCHSTONE and colophon are registered trademarks
of Simon & Schuster Inc.

Designed by Karolina Harris

Manufactured in the United States of America

10 9 8 7 6 5 4 3 2 1

Library of Congress Cataloging-in-Publication Data

Allman, William F.
    The stone age present: how evolution has shaped modern life—from sex, vio-
lence and language to emotions, morals, and communities/by William F. Allman.
        p.   cm.
    Includes bibliographical references (p. 255) and index.
    1. Genetic psychology. 2. Behavior evolution. 3. Human evolution.   I. Title.
BF701. A45   1994
155.7—dc20   94-18111

ISBN 0-671-89226-6
    0-684-80455-7 (pbk)

To Ryan Elizabeth and John Patrick

Time present and time past

Are both perhaps present in time future,

And time future contained in time past.

—T. S. ELIOT

Contents

Preface

---

*It is more important that a proposition be interesting than that it be true. . . . But
of course a true proposition is more apt to be interesting than a false one.*

—ALFRED NORTH WHITEHEAD

In many respects I have been writing this book all my life. I remember
as a child being fascinated by things that showed me where I was in relation to the world around me: I would scrutinize the diagrams in shop-
ping malls that showed an "X" affixed to a label that said *You are here;*
looking at maps while vacationing in Florida, I took a special pleasure
in the notion that as I walked along the shore that day, I was walking precisely along that black line between the orange peninsula and the light
blue sea.

Science, of course, is the most ambitious "you are here" diagram undertaken in history. Like the map in the mall, science attempts to show
us where we are within the context of the universe at large. Of course,
science is not the universe itself, nor is it the only way to get a feel for
what's out there. But science provides a profound sense of place, ar-
guably as powerful an experience as one can get from any other way of
knowing.

My goal in writing this book was to create a "you are here" diagram
for everybody's favorite subject—ourselves. The compartmentalization
of modern science, with its preoccupation with funding, politics and col-
liding philosophies, has created a scientific world that is a like a mall jam-

packed with scores of tiny boutiques, each shop specializing in its own wares and—sometimes—exhorting customers that it carries the only quality products in town. This mall of scientific knowledge is set in an exotic land where dozens of different languages are spoken, myriad customs are kept, and many of the "goods" are wrapped in plain paper with strange labels.

While the scientific shopkeepers are sometimes loath to admit it, at a fundamental level they are all dealing with the same wares. From the electronic flickering of a nerve cell in the brain to the intricate lacework of communal behaviors we call culture, human behavior can be fully understood only when all the ways we know the human self—from biology to psychology to anthropology—are seen as part of the same phenomenon, with each piece neatly interlocking into the other. Through most of my career I have explored the scientific realms of brain science, psychology, cognitive science, computers, animal behavior, linguistics, complex systems, archaeology, anthropology, and human evolution. But it has been only in the past few years that I have begun to realize that all of these topics are related to the same thing: the origins of why we behave as we do. The practical realities of modern science have pulled these areas of research farther and farther apart from each other. My goal was to put them back together again.

The result is a cascade of surprises about who we are and how we came to be: The way we behave today has its roots in the lives of our ancestors in the Stone Age; our huge brains arose primarily to cope with the enormous complexity of dealing with each other; the primary adaptation of our species is not hunting, toolmaking, or language, but our ability to cooperate. Ultimately, my research revealed that our species' remarkable evolutionary success is due to a trait that most of us take for granted, but that is rare among other animals on the planet: We make friends.

Drawing the connections between the disparate fields of the study of the human species would have been impossible without the help of the many scientists who welcomed me into their worlds, patiently explaining their ideas to me, often for hours and hours. These scientists include Randy White, Steve Pinker, Paul Bloom, Margo Wilson, Martin Daly,

David Buss, Don Symons, Robert Frank, Richard Leakey, Richard Klein, Donald Johanson, Napoleon Chagnon, Terrance D'Altroy, Craig Morris, Robert Tainter, and a host of others. Most important was the enormous help of Rick Potts, Robert Axelrod, Leda Cosmides, and John Tooby, whose ideas in many ways form the core of this book. Any errors in this book, of course, are my own doing.

I also thank the many editors during my career who took chances on stories that offered a different perspective on the human psyche and everyday life. Eric Schrier, Alan Hammond, Steve Budiansky, Wray Herbert, Roger Rosenblatt, Mike Ruby, and Merril McLouglin were instrumental in allowing me to flesh out some of the ideas in this book over the years. I am especially grateful to Christopher Ma for his unflagging enthusiasm for science writing. I also thank Sara Lippincott, Terry Monmaney, Paul Allman, Chris Nordlinger, and Josh Horwitz for their careful readings; Kristine Dahl for her urging me to do this book; and especially Robert Asahina for his keen editorial eye.

Most of all I would like to thank my wife, Patricia, without whom this book would not have been possible.

—W. F. Allman
March 1, 1994

Introduction

———

STONE AGE MIND

HOW EVOLUTION HAS SHAPED OUR MODERN-DAY BEHAVIOR

It is not for nothing that we have dubbed ourselves *Homo sapiens sapiens*, or "doubly wise." Buck naked, silent, and standing still, we fairly resemble any other of our larger primate cousins. Push the "play" button, however, and what is revealed is a psyche quite unlike any other intelligence in the world, perhaps the universe.

We are the most violent creatures on Earth—and the most cooperative. Our technological skills have allowed us to thrive from the Arctic tundra to the steamy mists of equatorial forests, and from the deserts of Africa to the concrete jungles of Manhattan Island. When we talk, make faces, wave our hands, jump up and down, blow into a saxophone, smear oils onto a canvas, scribble on a piece of parchment, or type into a computer, someone, somewhere, at a far corner of the globe or centuries into the future, nods in understanding. We imagine worlds that don't exist, laugh at things that have no meaning, and take delight in sights and sounds that have no mirror in nature. We extend our hands to total strangers, coddle our friends, bond with our lovers, worship our gods, make war against our enemies, and grieve over our dead. We are not the only animals to display these kinds of behaviors, of course. But other animals can muster only simple melodies; human behavior, on the other hand, is a hallelujah chorus—though the voices are not always in perfect harmony. Where did the maestro of all these modern-day behaviors—the human psyche—come from? And how do its ancient origins shape our modern minds today?

For centuries these questions have been the province of religion and philosophy. But now there is a revolution taking place in understanding the origins of the human psyche. A rapidly growing group of anthropologists, psychologists, linguists, philosophers, and other researchers are joining together to bring a fundamental part of the study of life on Earth—evolution—into the study of the life of the mind. These researchers are not the sociobiologists of old, who argued that human behavior is locked in the genes and governed by "survival of the fittest." But neither do they subscribe to the mainstream view of human behavior long held by many scholars, that the mind is almost exclusively the product of a person's surrounding culture, malleable as wet clay and as arbitrary as a shooting star.

Instead, these new researchers argue that over the eons of human evolution, the mind has been sculpted by natural selection to be adept at solving certain kinds of problems that were crucial for our ancient ancestors' survival—and that our modern-day minds continue to harbor that ancient legacy today. No one would dispute the fact that the lung and stomach were designed by evolution to breathe oxygen and digest food; why shouldn't the mind have evolved to do certain specialized tasks, too? These tasks are not necessarily those of modern-day city dwellers, who are only the tip of the iceberg of human evolution. Rather, the human mind evolved in response to the challenges faced by our ancient human ancestors, who for millennia lived in small groups of hunters and gatherers on the savanna.

The rich tapestry of behaviors that make up our modern everyday lives—our choice of a mate, our ability to live together in large groups, our love of music and concept of beauty, our anger in reaction to infidelity, our occasional hostility toward people who look different from ourselves—even the notorious male trait of never wanting to stop to ask for directions—all have deep-seated evolutionary roots that stretch back to the times when our ancient ancestors were struggling to meet the challenges of the world around them. These mental mechanisms are part of an ancient, Stone Age legacy that is still very much a part of the human psyche today, no matter what kind of culture or level of technological sophistication a person lives in—or whether the human is an ancient ancestor wandering the savanna or an astronaut orbiting the moon.

Though still in its infancy, this new approach to understanding the human psyche has already produced insights into human behavior. Studies show, for instance, that when it comes to sex and mating, men and women are as different psychologically as they are biologically. Around the world and across cultures, men show different preferences from women for what they desire in mates, for example, as well as how they attract partners and what makes them jealous, all of which hearken back to the times of our ancient ancestors. Other studies are revealing that far from being an all-purpose biological computer that can solve any problem equally well, the brain in fact has evolved to be more adept at reasoning about certain situations that were crucial to the lives of our ancient ancestors, such as avoiding danger, finding food, and catching people who are cheating on a social bargain. Still other studies reveal that human emotions, including our violent tempers, are not merely animalistic, irrational responses to situations, but rather are evolved mechanisms that help people conduct the challenging, intricate social relations that have characterized human life for eons.

Indeed, the most important finding of this new movement is the realization that one of the most complex, dangerous, and rewarding challenges facing our ancient ancestors was *each other.* Paleoanthropologists have long depicted our ancient human ancestors as a kind of Robinson Crusoe whose biggest task was finding food and shelter in a strange, hostile world. But, in fact, it wasn't the natural elements that posed the biggest challenge for Crusoe—it was the arrival of Man Friday. If Monday, Tuesday, Wednesday, and Thursday had turned up as well, notes psychologist Nicholas Humphrey, a pioneer in research on "social" intelligence, "then Crusoe would have had every need to keep his wits about him."

This view of the origins of the human psyche has come as paleoanthropologists and archaeologists are scouring Africa, Europe and Asia, discovering ancient fossils and artifacts that reveal a radically new picture of how the earliest humans went about their daily lives, from their eating habits to their social structure to their technology, language, and culture. The research suggests a view of our ancient human ancestors that contrasts with scientists' traditional image of the quaint, somewhat hairy version of our modern selves. The new findings reveal that, far from being predestined from the time our most ancient ancestors stepped down from the trees, the success of the human species was by no means inevitable. There have been no less than six different kinds of upright-walking, intelligent, apelike creatures that have walked the earth over the past 3 million years—several of whom were around as recently as 30,000 years ago.

The far-reaching message of these new researchers is that our incredible success as a species is not merely due to our upright walking, toolmaking skills, or language abilities—after all, many other creatures display inklings of these talents. Rather, the key to understanding our evolutionary success, as well as the unique combination of everyday behaviors that set us apart from every other living thing today, is our unique talents as social beings.

THE SOCIAL BRAIN

The incredibly sophisticated challenges of navigating a sea of other humans fueled the enormous expansion of the human brain size over the eons. It doesn't take many smarts to escape a tiger, find a safe place to sleep, or pull fruit from a tree—other animals do it all the time. But cutting a deal with a creature of equal mental abilities, whose motives may be nice or downright nasty, puts huge demands on the human psyche. Throughout our species' evolution, people's greatest challenge in their surroundings was to bond with, negotiate with, cajole, and sometimes threaten creatures who were just like themselves, maneuvering through a social environment that was sometimes treacherous, sometimes benign, but always demanding.

One of the specializations that the brain has evolved to help navigate social life is the unique human capacity for language. So important is language to human existence that the ability is hard-wired into the brain: Every human is born with an innate ability to learn any of the 5,000 languages spoken around the world today, or make up a new one if need be. Language is the ultimate social tool, enabling social relations to be taken out of the realm of the here and the now: A person can make promises or threats about what might happen in the future, be told about events in the past, and find out what happened while he or she was away. Language also makes each individual's behavior an integral part of the community at large: Word of mouth, rumor mills, referrals, and gossip all help create a social hotline that helps us determine whom to cozy up to, whom to butter up, and whom to avoid, without having to experience it first-hand.

Besides language, the mind contains a host of other specialized tools that have evolved to help grease the wheels of social relations. Parts of the human brain are customized, for instance, to respond to social signals such as smiles, frowns, and other expressions of emotions that form a universal communication system among humans. New experiments reveal that our minds reason better in social situations than when doing other kinds of tasks such as mathematics or logic problems. We also have the ability, unique in the animal kingdom, to make predictions about what other people are thinking. Studies are revealing that the physical, emotional, and societal factors that go into how we choose our mates, the difference between how men and women respond to jealousy and infidelity, our domestication of plants and animals, our ability to live in a complex society with leaders and laws, and even our appreciation of music and art all have a deep evolutionary legacy that is fueled by the demands of our intensely social life.

While the process of evolution is often characterized in terms of dog-eat-dog competition and the "survival of the fittest," this new research also reveals that, in fact, nice guys often finish first. Our species' success over the eons is due to the fact that we are the most *cooperative* creatures on Earth. We form vast, tightly knit, long-lasting bonds not just with members of our family—a trait common to many species—but with unrelated people as well. From the ancient savanna to the modern metropolis, humans have long forged enduring bonds with people who start out as strangers and go on to become partners in intimate relationships that are crucial to survival.

This quintessential human skill of forging relationships is no less important today, from the bedroom to the boardroom and from the far corners of the globe to the corner grocery store. Our lives are a constant making, unmaking, and redefining of relationships with others, weaving a vast social fabric of relations between brother and sister, parent and child, husband and wife, coworkers, neighbors, communities, leaders, and nations. Our very human propensity to view our world in social terms extends to nonhuman realms as well: We befriend cats, dogs, horses, and other animals; coax our cars and soothe our sailboats; curse the rain on a picnic day; and join in the most intimate of covenants with our gods.

The fact that our modern human behavior is influenced by our ancestors' evolutionary legacy does not mean that our modern behavior is necessarily desirable, inevitable, or unchangeable. Our craving for sweets, for instance, reflects the fact that in ancient times, sweet, ripe fruits were an excellent—and scarce—source of nutrition. In a modern world filled with an overabundance of candy and other goodies this evolutionary sweet tooth backfires. Most people, however, can successfully control their taste for sweets, and society has invented ways of fooling this evolutionary legacy with artificial sweeteners. Likewise, just as we have invented warm overcoats to cope with evolution's reluctance to outfit us with fur, the new research into the ancient origins of our modern behavior can help society create social inventions such as nutritional guidelines, laws, and community norms to help us cope with a modern world that is sometimes quite different from that of our ancestors.

In fact, the modern world can be seen as an unplanned "experiment" on our Stone Age minds. The ancient psychological legacies in our modern-day mind, while enormously useful in ancient times when our ancestors lived in tiny groups that hunted and gathered their food, sometimes are lost in the fast-food-filled, overcrowded, anonymous environment that characterizes modern society. Thousands of years ago, as the human population grew and people settled into towns, villages, and cities, they were suddenly forced to rely on strangers for food and mutual protection. The result was a new social environment with even more risks and potential pitfalls—and a new "cultural evolution" that resulted in laws, rituals, moral conventions, values, and state authority to knit together the social fabric of their community. At the same time, however, people living in complex societies became more vulnerable to exploitation by their rulers, group-against-group aggression and oppression, and the increasing environmental and health dangers that came with a sedentary life. These perils have led some scientists to question whether modern civilization, despite all its glory, is really all that good for us.

The question of how we as a species will survive the new challenges of modern society makes searching for the evolutionary origins of our modern-day psyches no mere intellectual exercise. There is more at stake in how we view the evolution of ourselves than the origins of, say, hedgehogs or crabapples. Like many origin stories of culture and lore around the world, the scientific stories we tell of who we are and how we came to be are more than mere information, for they serve as cautionary tales that remind us of our place as creatures of Nature. We may be the most intelligent, dangerous, and cooperative organisms to walk the earth, but we are not the final resting point of life's billion-year adventure on this planet. Our species is still an evolutionary work in progress; we are reinventing our lives every day and reshaping the world around us—and the final chapter is yet to be written. In this respect, understanding the ancient, evolutionary legacy that still sculpts our modern psyches is vitally important, for such knowledge undoubtedly will play a role in our species' survival in the future.

Chapter One

———

STORMING

THE CITADEL

No serious student of human behavior denies the potent influence of evolved biology upon our cultural lives. Our struggle is to figure out how biology affects us, not whether it does.

—STEPHEN JAY GOULD

Despite the rich, powerful complexity of human behavior—or perhaps because of it—most scientists who study why we behave the way we do ignore the fundamental question of where it all came from in the first place. Nature may have played a role in our ancestors evolving a large brain, goes the prevailing wisdom, but how that brain is nurtured by culture is more important in determining how we behave in our everyday lives. To most researchers in the social sciences, the brain is considered to be something of a *tabula rasa*, a general-purpose computer waiting to be programmed by culture. "Anthropologists have always assumed that evolution carried the human species up to the dawn of modern society and then left us there," says Harvard University anthropologist Irven DeVore. "After that, it is assumed that *culture* took over as the shaper of our behavior."

But in fact, *nature* and *nurture* are an inseparable blend of influences that work together to produce our behavior. A growing band of researchers are demonstrating that the bedrock of behaviors that make up the concerns of everyday life, such as sex, language, cooperation, and violence, all have been carved out by evolution over the eons, and that this Stone Age legacy continues to influence modern life today.

These researchers argue that it is impossible to understand how the mind works without asking the more fundamental, important evolutionary question, *What is the mind for?* "Imagine that you are an alien scientist who is sent to explore Earth, and you are confronted with a toaster for the first time," says Leda Cosmides of the University of California, who refers to herself and her like-minded colleagues as "evolutionary psychologists." "You may learn something about how the machine works by taking it apart or swinging it by the cord from a tree, or throwing it in the bathtub. But your efforts to find out how the toaster works would be aided enormously by the knowledge that the machine's function is to *make toast.*" Likewise, understanding our modern-day behavior can be helped enormously by the knowledge that, like the lungs, the stomach, or any other organ in the body, the mind evolved to perform certain tasks that were extremely important to our ancestors long ago. "Trying to understand the mind without first understanding *what it's for* is like trying to write all of Shakespeare's plays by having a roomful of chimps randomly typing away," says Cosmides.

The idea that, like any other animal on the Earth, humans display universal, species-wide, characteristic responses to situations involving food, sex, danger, or any other evolutionarily important part of our existence runs counter to the bedrock of mainstream social science. "It is an article of faith among anthropologists that a human being is an entity that passively learns culture—and that's the only thing everybody has in common," says anthropologist John Tooby, Cosmides' husband and also a pioneer in this new movement to understand the origins of the human psyche. "*Everything* else is assumed to be completely variable, because every culture is completely independent and different." But evolutionary psychologists argue that there is a fundamental human psychology that is part of everyone, around the world and across cultures. "When we see the tanks rolling into Tiananmen Square in China, or mass starvation in Africa, or children dying in some foreign land, we don't say, 'Gee, I wonder if these people are interpreting their experiences differently because of their different culture,' " says Tooby. "No: We see what is happening, and we say: 'I understand, as a human being, what's going on there.' And we understand because we have the same human nature that they do."

FALSE STARTS AND DEAD ENDS

One reason evolutionary theory has been banished from mainstream study of the mind is that past attempts to incorporate Darwin's ideas into the study of human behavior have proven disastrous. Biology had previously been used to justify racism, oppression, and exploitation long before Darwin came along, of course. But to a world steeped in colonialism at the turn of the nineteenth century, Darwin's formulation of evolution through natural selection was a powerfully attractive idea to some people. Many anthropological leaders of the day regarded technologically primitive cultures as vestiges of the distant past, destined to give way eventually to "more evolved" civilizations. In fact, it was the Social Darwinist Herbert Spencer, not Darwin, who coined the term "survival of the fittest," and used it to justify the ascension of the wealthy elite and the exploitation of the poor.

The notion that intelligence and other features of those people in so-called advanced civilizations were attributable to the "more evolved" status of certain races helped give rise to the eugenics movement, where prominent scholars sought to protect the purity of their races from genetic "mongrelization." Early in this century, a number of state laws in the United States authorized the sterilization of "undesirables" because, as one California statute puts it, these people had "an unstable state of the nerve system." Years before the rise of Nazism, German scholars discussed the progress of the eugenics movement in America as an exemplar of trying to maintain racial purity.

These ideas, which seem outrageous, even scandalous, today, serve as cautionary tales about the dangers of misusing science to support social and political goals. The Social Darwinists and the eugenics movement were no mere fringe groups but represented scholarly thinking of the time. Like so many of the bigots, tyrants, and social engineers of various political and social stripes who preceded them—and who exist in other forms today—Social Darwinists and eugenicists considered themselves well-intentioned people whose goal was to forge a new beginning for humankind, spread equality, free people from the shackles of their old ways of thinking, and so on. These ideas were given great weight in part because the ideology was advanced by leading scholars in academia—a tradition that continues today, albeit with different, but often no less tyrannical, philosophies of how to engineer society for its own good. Nor was the championing of evolutionary theory to support social theories confined to capitalist robber barons and fascists; Marx and Engels, too, regarded horticultural societies as primitive evolutionary "relics."

DISCONNECTING CULTURE

A strong voice against interpreting different societies around the world in terms of their evolutionary "status" arose early in the twentieth century. The anthropologist Franz Boas, founder of the "American School" of anthropological research, rejected the evolutionary trends prevalent in mainstream anthropology of the day, pointing out that people in so-called primitive cultures were just as mentally adept as those in the most sophisticated civilizations. Doggedly antiracist, Boas was reluctant to view a particular culture in terms of an ascending scale of evolutionary sophistication. Instead, Boas and his students, who included Margaret Mead, tried to understand each culture independently of comparisons to other cultures and civilizations. Concentrating on going out into the field and describing in detail a particular culture, Boas and his followers focused on outlining the differences among societies, not their similarities, and tried to understand a society's culture as an entity disconnected from the brains, biology, or race of the people who lived in it.

Boas began a deeply entrenched tradition in anthropological research that continues today in the work of leading anthropologists such as Marvin Harris, Marshall Sahlins, and Clifford Geertz. They and other modern anthropologists embody a research tradition that is based on three fundamental assumptions about the nature of culture: Culture is an independent entity and cannot be understood as a mere manifestation of biology or psychology. Culture is largely arbitrary, capable of taking an infinite variety of forms in a society. Lastly, it is culture—and not biology or genetics—that is the fundamental determinant of people's behavior.

By arguing that biology doesn't matter to culture, Boas and his follow-ers succeeded in overthrowing the erroneous, misguided, and racist no-tion that one race or culture is "more evolved" than another. Yet the anthropologists made the right gesture for the wrong reason, in effect throwing the Darwinian baby out with the bathwater. One culture is in-deed not "more evolved" than another—but not because cultures are arbitrary and infinitely variable and exist independently of biology, as the mainstream anthropologists believe. Rather, it is because, despite the myriad superficial differences, at a fundamental level the biology, psychology, and, ultimately, culture of all humans is *the same*. It is im-possible for one society to be culturally "more evolved" than another, because a society's culture is intimately connected to each person's bi-ology and psychology, and no human or group of humans is more bio-logically or psychologically evolved than another.

THE LEARNING BRAIN

With their assertion that human behavior cannot be fully understood without considering evolution, the evolutionary psychologists are offer-ing a radically new approach to understanding the human mind. Many people—scientists, science writers, and lay people alike—still believe in the misguided, antiquated, and wrong-headed "nature-nurture" dis-tinction that has been hashed over in social science for decades. That is, people still harbor the naive notion that either a behavior is "learned" through culture or socialization, or it is "fixed in the genes" as part of some sort of fundamental human nature that is immune to cultural in-fluences. Thus one hears sociobiologists talking about how our "nature" is kept in check by "culture," just as social scientists talk about how some stereotypical male and female behavior, for instance, arises solely be-cause people have been "socialized" by culture to act a certain way—with the implicit assumption that males and females just as easily could be socialized to act in exactly the opposite way.

These attitudes result from a profound misunderstanding of what genes do. The seemingly simple question, *Is this behavior in the genes?* in fact has several different meanings and answers. The first answer is that, at a basic level, *all* behavior is due to genes, because you can't grow a brain without genes, and you can't have behavior without a brain. Then there is the kind of question that people have in mind when they ask, Does Bob do this and Larry doesn't because Bob has the genes for it and Larry doesn't? Of course, part of the human genome varies from individual to individual, and that's why Bob may have a big nose and Larry has a small nose. But while these differences may be in the genes, they in fact have nothing to do with evolution. Rather, it is *having a nose in the first place* that serves an evolutionary function, and the genes for having a nose exist in every human being—whether that nose is large or small is largely irrelevant as far as evolution is concerned. Lastly, people sometimes use the phrase "in the genes" to describe a behavior that somehow overcomes a person's upbringing, or emerges as exactly the same regardless of a person's culture or environment, and so it is inevitable or predetermined. "We would say that there isn't any such thing as that," argues Tooby. "Genes are always in everything—but the environment is in everything, too. The whole nature/nurture debate boils down to the fact that we evolutionary psychologists say that there are hundreds of mechanisms in the mind that are specially designed to deal with various kinds of problems, while traditional social scientists say that there is some more general mechanism which is equally innate—though they don't use the word 'innate,' of course—that is some sort of general learning device."

One of the biggest misconceptions in the nature/nurture debate is that if a behavior has been crafted by *evolution*, then it is "predetermined" and "innate"; while behaviors that are *learned* are "flexible," "changeable"—and, ultimately, "improvable." Like the Social Darwinists of an earlier era, those who would like to use the findings of science for political and social purposes find the notion of an infinitely plastic mind attractive. In a world that is convulsed with inequality and strife, an infinitely pliable mind that can be "programmed" by society to behave in any fashion holds out the possibility of changing human behavior for the better through laws, teaching, or social revolution. Of course, one person's humane social engineering is another's Orwellian oppression: As the famed MIT linguist Noam Chomsky once pointed out, a *tabula rasa* model of the human mind is a "totalitarian's dream."

But, in fact, *learned* is not the opposite of *evolved*. The ability to learn is *also* an innate, evolved trait, which takes place within the biological confines of the brain. Just as the human physique unfolds according to a genetic plan, but is shaped by nutrition and physical stresses in each person's environment, the psyche, too, is carved out by a inseparable blend of evolution and environment. "*Everything*, from the most delicate nuance of Richard Strauss's last performance of Beethoven's Fifth Symphony to the presence of calcium salts in his bones at birth, is totally and to exactly the same extent genetically and environmentally codetermined," say Cosmides and Tooby. Our evolved mental mechanisms develop within the context of cultural influences that determine the exact nature of their expression. An evolved mental mechanism enables children to learn language, for instance, but it is culture that determines whether that language is English, Indonesian, Japanese, or whatever.

Since all human behavior depends fundamentally on the unique biology of the human brain, the real issue is not the old conundrum of "nature versus nurture," but rather, exactly what kind of "nature" do humans have that allows them to be cultural creatures? To many anthropologists and psychologists, the nature of the human mind considered to be "little more than a piece of soft wax onto which a person's surrounding culture makes its impression," says Cosmides. To these researchers, the mind is like an all-purpose computer that can be programmed to run any and all kinds of cultural "software," depending on what society tells it to learn. After all, a computer that is purchased for the purpose of processing payroll checks at a manufacturing plant, for example, is just as capable of analyzing medical records, helping design a sports car, or playing an unbeatable game of tic-tac-toe. Using the metaphor of the mind as a computer, it seemed appropriate to assume that the brain is simply another kind of general-purpose computer—one that is "programmed" by culture.

But there is a crucial flaw in using a computer as a metaphor for the human mind: The software program that enables a computer to do a payroll cannot automatically play tic-tac-toe as well—for each new task, a computer has to be *reprogrammed* with the proper rules and operations needed to perform the task. And while it might be a cinch to load various kinds of software programs into a silicon computer, the biological switches in a human brain can't simply be reset to perform some new task—the brain has to *learn* how to do these tasks on its own. Occasionally, people learn from explicit instructions, such as how to change a flat tire, draw letters, or make risotto. But in most instances people learn the zillions of things they need to know to get along in the world—from language to social skills to the very fact that they have an independent existence apart from trees, rocks, and other people—from observation and experience. For these important kinds of knowledge, people typically receive no explicit instructions. Thus the real issue for understanding human behavior is whether it is possible to create some kind of general-purpose computer that could *learn* to process payroll checks, play tic-tac-toe, and analyze medical records, *all from examples it sees in the world around it.* That kind of computer doesn't exist today.

It is unlikely that the human brain is a kind of general-purpose computer that is capable of learning any and all things equally. One reason is that the real world is full of many complex problems that require many different kinds of complex behaviors to solve. Just as there is no "general-purpose" organ in the human body that is capable of pumping blood, digesting food, and nourishing an embryo, it is just as unlikely that some sort of "general-purpose" brain could perform the entire repertoire of a person's behavior involved in finding food, choosing a mate, and selecting a place to live. "There is no such thing as a 'general problem solver,' " says evolutionary psychologist Don Symons, "because there is no such thing as a general problem."

Paradoxically, the enormous flexibility in how humans behave in everyday life suggests that the brain is an extremely specialized organ that is quite consistent in its responses. That's because the greater the number of potential responses there are, the more difficult it becomes to choose the right course of action. The brain cannot be some kind of general-purpose problem solver that merely sifts through the entire set of possible responses to a particular situation, weighing each one, and then picking the one that seems best, because there simply isn't enough time in the real world for such infinitely plastic behavior. "When a tiger bounds toward you, what should your response be?" ask Cosmides and Tooby. "Should you file your toenails? Do a cartwheel? Sing a song? Clearly, it is not the moment for the brain to be sifting through the pros and cons of a huge number of possible responses."

The simplistic notion that the source of all of our everyday behavior and culture lies merely in the brain's flexibility and learning powers begs all the interesting questions. As our ancestors evolved the behavioral plasticity that allowed them an expanded range of possible responses to a particular situation, there was also an increase in the number of potential responses that were dead wrong. The real question is, What kind of psychological mechanisms evolved in the brain that resulted in our ancestors' making the right choices?

That's precisely the question that Cosmides, Tooby, and their like-minded colleagues are beginning to answer. Cosmides and Tooby's insistence that evolution matters when thinking about the modern mind left them in an academic no-man's-land at the beginning of their careers. Cosmides was trained in psychology, but no one in that field was thinking about how the mind evolved. Tooby was trained in anthropology, but other than sociobiologists, no one was interested in the influence of evolution on the mind. "That's why we had to get married," jokes Cosmides. "When we met, we discovered that we were the only two people who thought the same way." Soon, however, the young couple's work began to receive recognition from the scientific community. Cosmides won the prestigious Behavioral Science Award—given to the best paper published in a given year by the American Association for the Advancement of Science—for research that grew out of her Harvard doctoral thesis. Tooby recently was given a President's Young Investigator Award from the National Science Foundation for his work. The couple's research has served as a focal point that has brought together a small group of like-minded researchers—such as Don Symons of the University of California at Santa Barbara, David Buss of the University of Michigan, and Martin Daly and Margo Wilson, a husband-and-wife team from McMaster University in Toronto—all of whom discovered that what they were doing was "evolutionary psychology," too. Now, Tooby and Cosmides are finding that their "no-man's-land" is growing increasingly crowded with researchers from a wide range of fields, from anthropology to psychology to linguistics to social science. "One of the reasons that evolutionary psychology took so long to get off the ground is that human behavior is enormously complicated, and so it requires an incredibly broad range of knowledge to fully understand," says Tooby. "Evolutionary psychology is a pivot between biology, psychology, and anthropology, and so you have to be familiar with the life-ways of hunter-gatherer societies, you have to be familiar with evolutionary biology, you have to know a lot about cognitive psychology, and genetics, and development, and so on. It is hard for any one person to encompass all of it. That's why you see couples like Leda and me, and Martin Daly and Margo Wilson, working together. We complement each other: My strengths lie in evolutionary biology, and Leda knows more about cognitive science."

THE WELL-TEMPERED PSYCHE

Researchers who claim that all of human behavior is due to mere "socialization" and "learning" must somehow overcome a problem that is frustratingly familiar to any parent: Why, despite the best efforts to instruct children to behave in certain ways, do kids appear to learn some kinds of things more readily than others? The answer to this question, argue Cosmides and Tooby, is that instead of being a general-purpose problem solver, the mind is actually a collection of special-purpose "mini-computers" that have been custom-programmed by evolution to learn about—and solve—specific kinds of problems. Just as a stereo system is made up of different devices such as speakers, a CD player, and a radio tuner, each with a specific function, the mind consists of largely independent modules that are specially tailored to perform specific tasks. Humans have innate predispositions to learn certain kinds of behavior, not because they are holdover instincts from our "animalistic" past, but because these mechanisms regularly produced beneficial behavior during the evolution of our ancient ancestors—and we inherited them. "As our species evolved, we didn't lose our 'instincts,' " says Cosmides. "Rather, we *proliferated* them."

Just like the lung or stomach, the mental "organs" that guide our behavior are part of *every* human's mental machinery. These mental modules are extremely complex, involving the interaction of hundreds if not thousands of genes, and were slowly hammered out by evolution as our ancient ancestors wandered the earth as small bands of hunters and gatherers. "Basically, we are talking about overall human design that transcends race or ethnicity," says Tooby. "It's not just a matter of ideological piety that makes us argue that people around the world are the same. There's actually a huge amount of biological and psychological evidence for this. Take a look around: You see two arms, two legs, and a head. And you could do a similar 'psychological' inventory: a jealousy mechanism, a 'love your kids' mechanism; a 'be attracted to fertile females' mechanism, and so on." There has not been enough time passed since the dawn of agriculture—which began a mere 10,000 years ago—for any new mental mechanisms to arise in response to our sedentary, urban, industrialized life. In other words, we may live in a modern metropolis, but in many respects we still harbor a Stone Age mind within our skulls.

Ironically, the evolutionary psychologists' model of the mind as being made up of hundreds of specialized mini-computers renders the champions of the general-purpose computer model of the mind as extremely reductionist—a criticism that has been rightly leveled at the gene-centered proponents of sociobiology as well. "The idea that our minds have hundreds of specialized mechanisms is too grimy and 'real world' for many researchers," says Tooby. "They like to think of the mind as being governed by a few simple general principles. But you can't approach the mind from the perspective of a physicist, looking for a process that can be reduced to a few elegant mechanisms. The mind is a bit like a car: You open up the hood, and there is all this incredibly complicated stuff. It looks hopelessly messy—if all you want to find is simple principles of how it works. But it is actually extraordinarily clean and precise—you just have to be willing to think not like a physicist but like an engineer."

The idea that the mind works on simple general principles is a result of the antiquated, almost Victorian notion that humans have a special evolution that is somehow different from that of other presumed lower animals. "There is this thinking that if other animals can do it, then it can't be real intelligence," says Cosmides. "So behaviors like eating, or finding a mate, or raising a child—all these things that other animals do—are thought to be simple. But when you actually try to work out the kinds of computational procedures you would need to do these so-called simple things, you find it's amazingly complicated. Take vision, for instance: For years people thought that the eyes were like little windows. You just open your eyes, and photons hit your retina, and you see. Then when they tried to do computer vision, they realized how complicated it really is; they couldn't get a computer to do these things that seemed so effortless."

But it is exactly when something seems effortless, says Cosmides, that researchers should suspect that there is something really complex and interesting underlying it in the mind. "People think that if something feels easy to do, the mechanisms behind it must be simple," she says. "We think that it is exactly the other way around: Things seem simple because evolution has crafted amazingly complicated mental machinery that is up to the task, and makes it *seem* easy."

MENTAL MODULES

Except for basic behaviors such as walking, smiling, and crying, specific human behaviors did not evolve. Instead, a number of psychological mechanisms evolved in the mind that, interacting with a person's surrounding environment, produce the cornucopia of behaviors that make up our everyday life. It is these psychological mechanisms, not a particular behavior in and of itself, that form the evolutionary legacy in our minds today. For example, a child may prefer to eat a candy bar instead of brussels sprouts, but the psychological mechanism in the brain that produces a craving for sweets has nothing to do with brussels sprouts or candy. The human sweet tooth evolved in response to fruits and honey, which were compact sources of high-quality nutrition. In modern humans, however, this same evolved psychological mechanism produces a new kind of behavior in response to a new kind of environmental stimulus—candy bars—fueling dental bills and a huge confectionery industry. Likewise, a species of tiny hermaphroditic fish has the remarkable ability to be able to turn into either the male or female sex, depending on the sex of the other fish around it. It is the *mechanism* that enables the fish to change its sex that is a product of evolution—not the actual *behavior* of changing into a male or female.

CHEATER DETECTORS

Evidence that evolution has crafted our minds to be specialized in certain tasks comes from Cosmides' remarkable discovery of a mental mechanism that is designed to detect "cheaters." Because they assume that the mind is an all-purpose computer, psychologists have long considered the mind's reasoning powers to be fundamentally the same no matter what the actual content of the problem a person is reasoning about. Whether it's apples and oranges or boxcars and baby buggies, the mind was assumed to reason in the same abstract way. But evolutionary theory predicts that the mind is specialized to be more adept at reasoning about some kinds of situations than others—particularly situations that are important to human survival, such as responding appropriately to danger signals, moving through the environment, having sex, and maintaining social relations.

Cosmides demonstrates the power of one such evolutionary mechanism with a simple puzzle: imagine that you have a new job as a clerical worker in a high school, and that part of your duties is to make sure the documents of the students have been properly labeled. Each of the student documents are supposed to be marked with a number and a letter. One categorization rule is: *If a student has a "D" rating, then his or her document must be labeled with a "3."* You suspect that the worker you replaced did not categorize the documents correctly. The following cards have information about the documents of four students at the high school. Each card represents one student; on one side of each card is the student's letter rating, and on the other side is the student's number code.

The cards read:

Which cards do you definitely need to turn over to ensure that none of the documents of any of these students violate the above categorization rule?

If you didn't answer, "Turn over the cards with *D* and 7 on them," you've got plenty of company. More than 90 percent of the people who try this puzzle get it wrong.

Now, try this one:

You are a Kaluame, a member of a Polynesian culture found only on Maku Island in the Pacific. The Kaluame have many strict laws which must be enforced, and the elders have entrusted you with enforcing them. To fail would disgrace you and your family.

Among the Kaluame, when a man marries, he gets a tattoo on his face; only married men have tattoos on their faces. A facial tattoo means that a man is married, an unmarked face means that a man is a bachelor.

Cassava root is a powerful aphrodisiac—it makes the man who eats it irresistible to women. Moreover, it is delicious and nutritious—and very scarce.

Unlike cassava root, molo nuts are very common, but they are poor eating—molo nuts taste bad, they are not very nutritious, and they have no other interesting "medicinal" properties.

Although everyone craves cassava root, eating it is a privilege that your people closely ration. You are a very sensual people, even without the aphrodisiac properties of cassava root, but you have very strict sexual mores. The elders strongly disapprove of sexual relations between unmarried people, and particularly distrust the motives and intentions of bachelors.

Therefore, the elders have made laws governing rationing privileges. The one you have been entrusted to enforce is as follows: "If a man eats cassava root, then he must have a tattoo on his face."

Cassava root is so powerful an aphrodisiac that many men are tempted to cheat on this law whenever the elders are not looking. The cards below have information about four young Kaluame men sitting in a temporary camp; there are no elders around. A tray filled with cassava root and molo nuts has just been left for them. Each card represents one man. One side of a card tells which food a man is eating, and the other side of the card tells whether or not the man has a tattoo on his face.

Your job is to catch men whose sexual desires might tempt them to break the law—if any get past you, you and your family will be disgraced.

Indicate only those card(s) you definitely need to turn over *to see if any of these Kaluame men are breaking the law*. The four cards read:

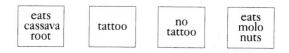

If you are like most people, you chose the cards that read *eats cassava root* and *no tattoo*. In fact, about 75 percent of the people taking this puzzle in a study by Cosmides got it right.

The intriguing thing about these two puzzles, says Cosmides, is that at a fundamental level they are exactly alike. In each puzzle, the logical task is the same: You have to pick the card that "proves" the rule—that the man who is eating cassava root has a tattoo on his face, or that the student's card with the *D* on one side of it has the number *3* on the other side. The tricky part for most people is knowing that they also have to check to make sure that there isn't a card that might "disprove" the rule. In the high school example, that means turning over the card with the *7* on it, to make sure that there *isn't* a *D* on the other side—which would break the rule. It also means that you must check the card that describes the dinner of the man with no tattoo, to make sure that he isn't breaking the rule that prohibits him from eating the prized food.

Why is it that people hardly ever realize they have to turn over the card with the *7* on it, and nearly everyone realizes that they have to turn over the card to see what the man with no tattoo is eating?

The reason for this, says Cosmides, has nothing to do with the Kaluame people, cassava root, high school students, or documents. But it has everything to do with the evolution of the human psyche. The high school version of the problem simply asks you to match a letter with a number—a highly abstract task that would have had little relevance to our ancient ancestors. The version of the problem involving cassava root—even though it is set in an exotic, fictional locale that is unfamiliar to everyone—nevertheless calls upon the mind to do a task that has been fundamental to human social life around the world for eons: catching people who are not keeping up their end of a social bargain. The cassava root problem is far easier to solve, says Cosmides, because our minds have been customized by evolution to perform precisely these kinds of tasks, which involve what has long been the most challenging part of our ancestors' environment, *each other.* "Many of the most important problems our ancient ancestors had to face were social," she says. "They needed to know how to cooperate, how to respond to threats, how to participate in coalitions, how to respond to sexual infidelity, and so on. The result is that the human mind contains a number of specific mechanisms that were specially designed by evolution for processing information about the social world. One of these mechanisms is a 'cheater detector.' "

IN THE CITADEL

With their finding that the mind's reasoning powers are a product of the repertoire of "mini-computers" crafted by evolution, Cosmides and Tooby are revolutionizing how psychologists think about our mind and behavior. "There are some psychologists who are willing to concede that certain aspects of the human psyche—such as the ability to use language—may be the result of a specialized mental 'organ,' " says Tooby. "But they still believe that the rest of the mind is a 'general problem solver'—they simply banish language to a mental 'ghetto.' But we're saying, 'The hell with the ghetto, we're going to the very center of the territory, to what people think is the heart of the mind: *reasoning.*' It's the ultimate cognitive activity—and we're showing that it also has an evolved, modular, domain-specific component, and that's really trouble. They can't banish us out to a ghetto, because we are at the center of the citadel."

For years psychologists have known that people perform better if an abstract problem is put into more familiar terms such as reasoning about food or transportation. But while people tend to do better when the problem is offered in realistic settings, psychologists have long been perplexed over the fact that people's performance is erratic, with some familiar scenarios evoking more accurate responses than others. Cosmides' experiment reveals, however, that it is not familiarity with the subject matter that triggers better reasoning but the type of transaction that is taking place. "That's why our approach is so contrary to mainstream psychology," says Cosmides. "Most psychologists are looking for a general problem solver, and so for them, the finding that people perform differently on tasks whose content is different is annoying. But for us, *content is everything.*"

Cosmides reasoned that, given the importance of social relations throughout human evolution, people should perform better on a logic problem when it is presented in scenarios depicting the interactions of a social contract, rather than being purely descriptive—regardless of how familiar the setting of the problem. One central principle of cooperative social contracts is the notion "You scratch my back, and I'll scratch yours." If a person takes a benefit from someone, he or she must return a benefit to that person. "Cooperation can't evolve without 'cheater detectors,' " says Cosmides. "If you always give benefits to others, without getting anything in return, you wipe yourself out." The result is that evolution would have designed the human brain to be acutely sensitive toward "cheaters."

To make sure that the people were responding to the social setting of the Kaluame problem and not some other factor, Cosmides also tested subjects with a similar version of the problem. In this case, the island setting and the identical "rule" to be tested were the same, but the scenario described was one where no social contract was involved: The cassava root and molo nuts foods were described as equally nutritious, but the nuts grew on one side of the island and the root on the other, and men who lived on one side of the island had tattoos and those who lived on the other side did not. In this test, Cosmides found her subjects chose the logically correct answer only 25 percent of the time.

Furthermore, Cosmides found that people persist in reasoning along the lines of social contracts even when doing so is strictly illogical. She gave people the same "social contract" scenario as in the first test, but the rule was reversed to say, "If a man has a tattoo on his face, then he eats cassava root." The same four cards were given, reading: *eats cassava root, no tattoo, eats molo nuts,* and *tattoo.* Because the rule was reversed, this time the "logically correct" response is to turn over the cards reading *tattoo* and *eats molo nuts.* Cosmides' subjects, however, nevertheless chose the cards *eats cassava roots* and *no tattoo* nearly 70 percent of the time anyway. While a mathematician would point out that this response is logically incorrect, these responses are nevertheless the correct behavior in the important business of catching cheaters. "People may be acting illogically from the point of view of a logistician," says Cosmides. "But they are doing the absolutely correct thing from the evolutionary point of view. If you used the rules of logic, you might get the right answer, but that would also mean that you are a very poorly designed human being." People reason in accordance with enforcing social contracts, not logic.

In fact, what the mind looks for in deciding who is "cheating" on a social contract depends on the perspective of the person on the lookout. In one study, for instance, people were told that they were the head of a company and given the rule: "If an employee works for ten years, then he or she must receive a full pension." The subjects were then asked which of four cards they need to turn over to prove that the rule had not been violated. The four cards represented four employees, and read *worked ten years, no pension, worked less than ten years,* and *receives pension.* When people were told they were the owner of the company, the majority chose the cards *worked less than ten years,* and *receives pension.*

But when subjects were told that they were employees of the company—not the owner—their responses were the opposite, even though the rest of the problem was exactly the same: The subjects chose *worked ten years,* and *no pension.* The different responses, says Cosmides, arise because, to an employer, cheating is defined as those employees who take a pension without having worked the required number of years. To an employee, however, cheating is defined as having worked the required time and receiving no benefits from the employer.

ANCIENT STATISTICIANS

Cosmides and Tooby have found further evidence that the reasoning power of the human mind has been sculpted by the evolution of our ancestors—in the way we think about probability. "Psychologists have always assumed that people are bad statisticians," says Cosmides. "And, in fact, experiments seem to show this." In one such experiment, for instance, people are told that the prevalence of disease in a population is one in a thousand. They are then told that there is a test to diagnose the presence of this disease, and it has a "false-positive" rate—that is, the test incorrectly indicates the presence of the disease—of 5 percent. The question is, what are the chances that a person who tests positive for the disease actually has it? "This test was given to a bunch of experts, including members of the faculty at Harvard Medical School, and they completely bonked out on it," says Cosmides. "Less than 20 percent gave the right answer."

But there is something fishy about this problem, says Cosmides. It might be that it is the way the problem is presented, not the reasoning task itself, that poses the most difficulty. "People have a very difficult time dealing with probabilistic information," she says. "When a physician says you have a 2 percent chance of getting breast cancer, what does that mean? Does it mean that each of us has a 2 percent propensity for getting breast cancer, or that out of 100 people, there is a 100 percent chance that 98 of them will be healthy and a 100 percent chance that 2 of them will have breast cancer?"

It is unlikely that the brain evolved to do this kind of statistics, she says. "If you are a hunter-gatherer wandering around the landscape, you are not encountering these numerical abstractions about probabilities— you are encountering real frequencies in the real world. A woman gathering food for the day is not going to think to herself, 'Hmmm, there is only a 10 percent probability that there will be fruit on that tree over by the lake.' She's going to say to herself, 'I've been there ten times before, and I've found fruit only once.' It makes sense from an evolutionary psychological perspective that if the mind had a psychological mechanism for doing statistics, it would be based on frequency-type inputs and give frequency-type outputs. People in fact may be very good at reasoning about statistics—but only if you give them the problem in a format that the human mind evolved to deal with; that is, in frequencies."

To test this idea, Cosmides and Tooby recast the medical puzzle in terms of frequencies: One of every thousand Americans has disease *x*. There is a test to diagnose the symptoms of disease *x*, and everyone who has the disease tests positive for it. But because the test is not perfect, some healthy people also test positive for it. Specifically, out of every

1,000 people who are perfectly healthy, 50 of them test positive for the disease. Imagine that a lottery were conducted, and 1,000 people were randomly selected and given the test. How many people who test positive for the disease actually have it? "It turns out that when you think of the problem this way, it becomes far simpler," says Cosmides. "You've got a population of 1,000 people, and since the test isn't perfect, 50 of the people will be healthy but test positive for the disease anyway. And since 1 out of 1,000 people actually has the disease, that person is going to test positive, too. So in all, 51 people will test positive for the disease. When you ask how many people who test positive for the disease actually have the disease, it is 1 out of 51." Cosmides and Tooby found that rephrasing the problem has dramatic results in how well people do on it: Given the original version of the problem, only 12 percent of the people tested got it right. But on the rephrased version of the problem, 76 percent of the people got it right. With one group of subjects, Cosmides and Tooby forced them to think in terms of frequencies by giving them a piece of paper with little squares on it, telling them that each square represents a person, and asking them to circle the people who test positive for the disease and color in those who actually have it. In that version of the test, 92 percent of the people got it right.

SEX ORGANS IN THE MIND

Evolutionary psychology suggests that just as men and women have a few different biological organs that help them carry out the different biological roles they play in life, some evolved mental mechanisms will be different in men and women, too. For instance, several ingenious studies have found evidence that men and women have slightly different evolved mental specializations for how they deal spatially with their environment. Males have long been thought to simply have better spatial skills, because they tend to perform better in spatial tests such as trying to mentally construct a three-dimensional object from a flat, two-dimensional outline. But evolutionary psychologists argue that males' spatial skills are not superior to those of females, but merely different—and that this difference may have its roots in how our ancient ancestors went about obtaining their food.

In one experiment, men and women who had volunteered for a study were asked to wait in a small office while the apparatus was being set up. The office was cluttered with objects such as books, pictures, coffee cups, etc., and subjects waited some ten minutes before being called to another room. When they finally were called in, the subjects found to their surprise that the "experiment" was for them to try to recall the various objects, and their location, in the room they had just been waiting in. Contrary to conventional wisdom, in this particular kind of spatial task women far outperformed men, from remembering the number and type of objects in the room to pinpointing their location in the office— even in one group of men and women where the ruse was dropped and they were told the exact nature of the test before they went into the waiting room. In another study, men and women learned to maneuver a spot of light through a complex maze that appeared on a computer screen and to go through a life-size maze. The researchers found that women navigated by relying on environmental cues such as a crack in the wall or a different pattern of flooring: When analogous "landmarks" were removed from the computerized maze, for example, women got lost more than men. And when men and women were shown photographs of various junctions in the university's vast underground tunnel system, the women were better at identifying the location in the tunnel system where the picture was taken. Conversely, more men than women got lost when the shape of the computerized maze was distorted; on the other hand twice as many men than women correctly picked out a diagram that represented the overall shape of the life-size maze. These findings suggest that, unlike women, men appear to navigate not so much by environmental cues as by creating a mental "map" of the overall pattern of their journey—noting that they should go straight ahead for a stretch and then take their third left, for instance. Women, on the other hand, tend to rely on landmarks and cues from the environment.

Males and females differ in their spatial ability in modern times because of the special tasks their brains evolved to solve in the past. During the vast majority of time that humans walked the earth, people have lived in small groups that sustained themselves through hunting and gathering. In these groups there was a division of labor, with men doing most of the hunting and women doing most of the gathering (with gathering, in fact, providing most of the food for the group on a typical day). Women's primary role as gatherers resulted in the evolution of a mind that was adept at keeping track of what kind of objects were in which locations in the surrounding area—remembering that there was a tree bearing fruit, for instance, near a certain part of a particular stream. Such spatial skills would have been vital for a gatherer whose "quarry" is widely dispersed but stationary. Men, on the other hand, had to struggle with the logistics of hunting animals who were constantly on the move, sometimes in unfamiliar territory, leading to evolutionary pressures for a very different set of spatial skills involving reasoning about spatial relationships. This difference in men's and women's mental mechanisms for spatial reasoning may lie behind the bane of modern couples when they travel together by car: a man's unwavering reluctance to ask directions when he is lost. Since men rely more on their sense of movement and direction rather than landmarks when traveling, they are less prone to notice unfamiliar sights that would suggest that they are lost—as compared to a female traveling companion, whose ability to spot landmarks would make her more adept at noticing sooner that they are in an unfamiliar setting.

The findings that men and women have fundamentally different spatial abilities—not simply better or worse because of adverse cultural "socialization"—touches the essence of the evolutionary psychologists' impact on the understanding of human behavior. Because psychologists have long regarded the mind as a "general problem solver," differences between the performance of men and women on psychological tests have been characterized as *deficiencies*, typically thought to be the result of cultural influences. But in the evolutionary psychologists' view of the mind as having been customized by evolution, the minds of males and females are different in subtle ways because their evolutionary history is subtly different. "An evolutionary perspective can help eliminate biases against women," says Cosmides. "For fifty years we've had a bunch of male scientists testing spatial cognition. Since mainstream social science doesn't have a real theory about what the mind is designed for, these scientists simply follow their own instincts on what constitutes a spatial 'problem'—and since they are men, they define spatial cognition in a geometric sense. Of course, if it had been a bunch of women psychologists studying spatial cognition, then they would have designed the tests that had to do with the spatial *location* of things, and we would have been told that it is *men* who have the terrible spatial cognition. But when you have a principled theory about the evolution of the mind, you can avoid sex bias and ask, What kinds of mental mechanisms would you expect to find to help men and women solve these different tasks? And you'd expect to find sex differences—not 'better' or 'worse,' just different—when their problems were different, and you would not find sex differences when their problems were the same."

AN UNPLANNED EXPERIMENT

Because our mental organs evolved during the times of our ancient ancestors, the modern, industrialized, overcrowded world often presents these evolutionary mechanisms with novel circumstances that would never have been encountered by our ancestors. Because evolution operates on the psychological mechanisms that produce behavior—but not behavior itself—modern humans who are confronted with novel environmental inputs in the modern world may act in ways that directly go against the purpose for which the mechanism evolved.

This point is underscored by a study of incest avoidance by children who were raised on kibbutzes in Israel. In a typical kibbutz the children of the same age from different families are raised together in a single group. Examining the marriage records of adults who were raised in kibbutzes years ago, researchers found that the incidence of marriage among one group of adults who were raised together starting from a very young age was far below that of another group of adults who joined the kibbutz later in their childhood. Even though the people from both groups were unrelated, those who were raised together from a very young age didn't marry.

The research is strong evidence that the avoidance of incest, which appears in societies worldwide, is the result of an interaction between an ancient evolved psychological mechanism and the modern environmental cues around it. Incest is a maladaptive behavior, evolutionarily speaking, because if two close relatives conceive a child there is a greater likelihood that some rare, inherited genetic disorder might appear. Yet the individuals in the kibbutz study were unrelated, and so their behavior could not have resulted from some kind of biological mandate that tallies up the genetic pros and cons of a particular behavior with the goal of furthering one's genetic input into the next generation. Nor was there any "learned" cultural prohibition against marrying one's kibbutz partner—in fact, it was hoped by many parents that their children would marry someone from the same group. The study suggests that incest avoidance is the result of an evolved mental "mechanism" that takes the form of *don't be sexually attracted to people with whom you are raised from a very young age.* This mechanism is generally effective because, in the times of our ancient ancestors, children who were raised together starting from a young age *were* brothers or sisters or close cousins. It is only in rare "environments" such as a modern kibbutz that this evolved mechanism misfires and keeps unrelated adults from marrying. This mechanism misfires in reverse, too: In the rare cases where brothers and sisters are separated at birth and reunited as adults, they sometimes fall in love.

This incest-avoidance mental mechanism takes environmental inputs—the presence of other young children in close proximity—and acts on this input to produce behavior—inhibiting sexual attraction toward those people. The existence of such an evolved mental mechanism suggests that incest should occur far more frequently among unrelated individuals who join a family later in their lives, such as among stepchildren or stepparents and their children—and this is the case. Since these evolved mechanisms depend on environmental cues to develop, the mechanisms will not appear as some sort of invariant, monolithic behavior that is immune to cultural influences but will exist in many different shades and variations. The fact that these mechanisms do not *automatically* produce a specific behavior, regardless of the environmental setting, is why there are laws and cultural taboos against incest.

BEYOND SOCIOBIOLOGY

With their focus on how evolution has sculpted the mind, evolutionary psychologists are moving beyond the sociobiological argument that the fundamental shaper of human behavior is some sort of innate, biological drive to put one's genes into the next generation. "Many sociobiologists have this view of people as *fitness maximizers*," says Tooby. "They assume that since evolutionary biology says, 'We all evolved to propagate genes,' the purpose of humans is to propagate genes. They believe that beneath all of our complicated human behaviors there is an underlying hidden logic of 'gene propagation.' So when you are being nice to your child, they say, all you are *really* doing is selfishly trying to propagate your own genes. A lot of sociobiological work carries this cynical interpretation of human behavior—a view of the world for which sociobiologists have been rightly criticized. The problem is that sociobiologists confuse the mechanisms of the mind with the *process* that built the mind, and in fact these are two separate things. Evolutionary biology is not a theory of human nature. Rather, it is a theory for how human nature *came to be*—and a useful tool for discovering what human nature actually is. A mother really does love her child—it's not that somewhere deep inside her mind there is a selfish motive to spread her genes. In fact, it's really the other way around: Human beings love their children because those ancestors who loved their children had more surviving children, and we're descended from them and not the others who didn't love their kids. So in the 'grand evolutionary biological' sense, we love our children because of genes. But in the real sense of *Why do you love your kids?* you love them because it is part of your human nature that evolved as part of our ancestors' brain mechanisms. There is nothing in those brain mechanisms that says *That kid has your genes; he's propagating your genes, and so you should love him.*"

Cosmides and Tooby argue that the key to understanding how evolution influences our everyday behavior is not the *gene* but the *mind*. The human mind contains psychological mechanisms that interact with the environment to achieve goals such as finding food that tastes good, having sex with attractive mates, and catching cheaters. These goals are roughly correlated with evolutionary "goals" such as getting ample nutrition, reproducing healthy children, and not allowing others to compete through deception. But it is satisfying these proximate, psychological goals—not fulfilling the genetic "goals" of evolution itself—that is the prime motivator of our everyday life. "There is *proximate* causation and *ultimate* causation," says Cosmides. "The ultimate cause of why people like to eat nutritious food—and not gravel and broken glass—is that our ancestors had food preference mechanisms, and those whose mechanisms led to their having more children thrived, and we inherited their mental mechanisms. But the real reason we eat certain foods is because *they taste good!*"

This subtle, but crucial, difference between evolutionary psychology and sociobiology can be illustrated by a simple example: "When a woman puts salt on her eggs for breakfast," says Cosmides, "a sociobiologist would attempt to explain such behavior by trying to determine how the presence of salt on her eggs increases the number and fitness of her offspring in the future. An evolutionary psychologist, on the other hand, asks what kinds of psychological mechanisms for craving salt might have evolved in our ancient human ancestors." The difference in explanatory power between the two approaches lies in the fact that even when modern medicine suggests that too much salt may well be harmful—and so may *not* increase her evolutionary fitness—the news does not diminish the importance of evolution in explaining our everyday behavior.

That's because our everyday behavior reflects the lives of our ancient ancestors in the past—not our attempt to get our genes into the future. "A fast-food restaurant is a little monument to the diet of our ancient ancestors," says Cosmides. "Fast food has all the things that were very hard to get for our ancestors—such as salt, sugar, and fat—and so having an appetite for these things was very important back then. Sugar is in fruits, and you can't get too much of them; wild animals don't have a lot of fat on them, so having a taste for fat makes you want to get meat, and you could never get enough of that, either. By the same token, our ancestors didn't need an appetite that would cause them to go out in search of fiber, because every plant they ate gave them tons of fiber. Unfortunately, we have inherited those evolved tastes—even though we now have all the fat and sugar we want. But since our ancestors could never get enough of this stuff, we don't have an evolved mental mechanism that says, 'Aha, I've had enough sugar and fat.' We don't want it any less, even though we know the health consequences of it, because it still tastes good." The fact that we harbor a Stone Age appetite in a modern-day world does not doom us to forever overindulge in sugar, fat, and salt. The very knowledge that we possess such a predisposition helps us change our eating behavior in the present, and inspire us to invent salt, sugar, and fat substitutes, for instance.

As the research of Cosmides, Tooby, and other evolutionary psychologists demonstrates, merely having an all-purpose problem solver for a mind was not enough for our ancestors to get by in their complex and dangerous world. Instead, evolution carved the human brain into a collection of specialized mental mechanisms that helped our ancient ancestors meet the challenges of their everyday lives, whether it was obtaining food, finding shelter, or escaping from predators.

The biggest of these challenges, however, was dealing with the most treacherous, dangerous animal in their world: each other. It was the dangers—and payoffs—of navigating the social world that triggered the evolutionary ballooning of the human brain.

Chapter Two

THE SOCIAL

BRAIN

One man that has a mind and knows it can always beat
ten men who haven't and don't.

—GEORGE BERNARD SHAW

It was a balmy day in Kenya's Amboseli National Park, and Robert Seyfarth and Dorothy Cheney were searching for signs of humanity. They had rigged up a loudspeaker behind some bushes, and were broadcasting various shrieks, chutters, and barks they had previously recorded while observing the large group of vervet monkeys that make Amboseli their home. A husband-and-wife team of primatologists from the University of Pennsylvania, Cheney and Seyfarth were replaying a recording of the distress call of a young monkey they called Emerson. Emerson was the son of a female monkey dubbed Teapot Dome; she had wandered off with two other females to find food and was out of sight of her son. The researchers had intended only to see whether Teapot Dome could recognize her son's distress call even when she couldn't see him. But when Cheney and Seyfarth later examined their videotapes of the experiment, they discovered a subtle clue to the origins of the human mind. Emerson's screams indeed did cause Teapot Dome to look with concern in the direction of the speaker; but more revealing was the reaction of her two companions: They immediately turned and looked expectantly at the mother. "It was almost as if they were thinking, 'That scream goes with Emerson, and Emerson goes with Teapot Dome. What's she going to do about it?' " say the researchers.

The fact that our evolutionary cousins take a keen interest in the personal affairs of each other perhaps should come as no surprise to a species that thrives on gossip, newspaper tabloids, and TV soap operas. We are merely the most intelligent and social members of the most intelligent and social family of animals in the world: the primates. Cheney and Seyfarth may be probing the mind of a little monkey in Africa, but their research promises ultimately to shed light on the bigger question of the origins of the human psyche. Just as you can learn a lot about the origins of the human hand by studying the hand of a monkey, the behavior of nonhuman primates can help shed light on the origins of human language, cognition, and self-awareness. Primates can help us discover not only how human intelligence evolved but *why*.

Ironically, what makes us uniquely human is a trait we share with many of our nonhuman evolutionary relatives: *social* intelligence. Many primates live in societies marked by a swirl of social alliances, intrigue, and deception; a game with multiple players who must make their way through a thick social network. "Watching monkeys, one is tempted to treat them like tiny humans," say Cheney and Seyfarth. "Not only because they look rather like us but also because features of their social organization—close bonds among kin and status striving, for example—look like simplified versions of our own."

Humans, of course, are the quintessential "deal-making primate," taking these typical primate behaviors and amplifying them into an incredible array of complex interactions. What enables us to be the champions of the social circuit is the three-pound lump of nerve tissue lying within our skulls: the human brain. The brain is the crucible of behavior. It is there that the legacy of genes and shapings of experience are melded to create our thoughts and actions. In the brain, the mind is made flesh: The ephemera of thoughts, memories, and experiences are put into the hard, wet, biological reality of nerves, synapses, and neurotransmitters. The brain recasts itself with our every new experience, constantly changing its physical form in response to its surroundings, reweaving its biological warp and woof with each new memory. The seat of the self, the brain is also the soul of our species. So enamored are we of our mind that we have named ourselves after our brain—*Homo sapiens sapiens,* the "doubly wise" human. This outsized gray-and-white mush makes us the oddballs of neurobiology charts of the animal kingdom—just as the behavior that this organ produces has made our species sometimes seem like strangers in the natural world.

GETTING A SWELLED HEAD

Over the course of human evolution, the brain of our ancient ancestors nearly tripled in size, from about the size of a baseball in our 2.5 million-year-old ancestor *Homo habilis* to its modern-day bulk—about the size of your two fists put side by side. Part of this increase is simply the result of the fact that the human body, too, has grown bigger over the eons. The brains of elephants and whales, for instance, are bigger than those of humans, because these animals' bodies are bigger. Yet when our brain size is considered in relation to our body size, it once again becomes the leviathan of the animal kingdom: It is nearly twice as large, per body size, as the brain of an elephant and three times larger than what might be predicted for a typical primate of our size.

Bigger brains are not necessarily better brains. The average size of the modern human brain is about 1,350 grams; the novelist Ivan Sergeyevich Turgenev had a brain that weighed over 2,000 grams; the writer Anatole France reportedly had a brain of just over 1,000 grams. Some modern human brains are smaller than the largest brains estimated for the ancient human ancestor *H. erectus;* and another human ancestor—the Neanderthals—had an average brain size estimated to be slightly larger than that of modern-day humans—about 1,400 grams.

The modern human brain is not only bigger than those of most other animals, it is organized differently. You can't produce a human brain simply by taking the brain of a chimpanzee, for instance, and expanding it. In a chimp brain the neocortex—the part of the brain involved with higher learning, reasoning and perception—forms about 75 percent of its total bulk; in a human brain the neocortex makes up about 85 percent of the total mass. The human brain is also different from that of other species in that each hemisphere has become a specialist: In humans, the left half of the brain typically is responsible for language, symbolic reasoning, and dexterity, while the right half handles spatial perception, emotion, and music appreciation. In other words, just as modern society derives its power from people being specialized into farmers, craftspeople, and soldiers, etc., the brain appears to have increased its power by dividing up the work among various parts of its tangled neural structure.

A shining example of the incredible complexity that can arise in living tissue, the brain is no less a product of evolution than any other complex, specialized animal trait such as a bat's ability to navigate using echoes or a spider's ability to spin a silky web. Nor is the human brain's great increase in size some sort of evolutionary anomaly: For the brain to have undergone its huge ballooning over the past 2.5 million years of human evolution, it need only have grown a millionth of one percent per generation. And it is not all *that* unusual for humans to have large brains, for having a large brain is a trait that appeared in the primate family long before the arrival of humans. The brain of the ancient ancestor of apes, monkeys, and humans underwent a 65 percent increase in brain size some 60 million years ago, and kept expanding.

LOOKING FOR THE TRIGGER

Of course, having a big brain comes with a price: Like a V-8 automobile engine, the human brain is a gas guzzler. In adult humans the brain consumes 20 percent of the energy of the body at rest—as opposed to only 9 percent in the average chimp. Also like a V-8, big brains are sensitive to overheating: A rise of only four degrees in body temperature can lead to fuzzy thinking in adults and convulsions in children. Another problem created by our ancestors' ballooning brain is that a big brain means a big head, which makes giving birth dangerous to both mother and child.

In this respect, brain size is not the only feature that distinguishes our species from the rest of our primate cousins: Humans have one of the shortest gestation periods, relative to their overall life spans, of all the primates. If human mothers followed the gestation pattern typical of other primates, they would give birth 18 months after conceiving a child.

In fact, at birth, the infant brain is only about a quarter of its adult size. It begins its biggest growth spurt only after a child is born. The human brain is larger than that of other primates precisely because it keeps up this childlike growth spurt long after the brains of other primate infants have stopped growing.

This prolonged, infantlike growth rate in humans—and the big brain that came with it—may have evolved through mutations in the genes that control our overall development: Unlike other primates, our "juvenile" form is extended into adulthood, along with our juvenile rate of brain growth. In fact, human adults resemble infant apes, whose bulging foreheads and flat faces eventually develop into the jutting-chinned, sloped-forehead shape of adult apes.

Given all the trouble to a body that a big brain causes, it is clear that it was no mere accident that big brains arose in our ancestors. The intelligence tests our ancestors faced in everyday life must have been formidable, because solving those problems made the smarts that came with having a large brain outweigh the enormous biological disadvantages that brain placed on their bodies.

Over the years researchers have proposed various scenarios that might have "triggered" the explosion of the human brain. The first and most obvious was walking upright. It seemed plausible to assume that once our ancient ancestors began to stand on two legs some 3.5 million years ago, then the appearance of other human traits such as toolmaking and large brains quickly followed, creating a feedback cycle that led to a larger brain.

LUCY

This scenario began to fall apart with the famed discovery of the fossil of a diminutive, ancient human ancestor, popularly known as "Lucy." Dating back to nearly 3.5 million years ago, Lucy and her kin represent the oldest known direct human ancestor. Her skeleton reveals that she walked upright—in fact, most researchers agree that her kind, known by the scientific name *Australopithecus afarensis*, probably made a series of remarkably humanlike footprints that were discovered frozen in rock-hard volcanic ash in 1979. Dating back 3.7 million years, the footprints clearly show the 80-foot path of several hominids as they walked along the savanna, paused slightly, then pressed on. The structure of Lucy's legs, knees, and pelvis reveals that she was well adapted for an upright stance.

As strikingly humanlike as Lucy was from the neck down, however, she and her contemporaries were remarkably *un*humanlike from the neck up: The hominid had a tiny, apelike brain, large canine teeth, and a protruding face—an ape's head on a human body. There is no evidence that Lucy and her kin made stone tools, harnessed fire, built shelters, retained long-term mates, or focused their lives around a central home base. All evidence suggests that Lucy's brain, and behavior, was very apelike.

Because the first big leap in brain size in the human lineage appeared along with the first sign of stone tools some 2.5 million years ago, other scholars looked at toolmaking as the key to brain growth in the hominid line. As did filmmakers: The opening scene of Stanley Kubrick's *2001: A Space Odyssey* shows an ape wielding a piece of bone in a moment of simian epiphany. But the idea that toolmaking spurred the leap in brain size began to unravel, too, as researchers discovered little evidence of organized hunting or other signs of big brainpower among the remains of the first stone toolmakers, who are known as *H. habilis* or "handy man." Besides, if the dawn of the human race did occur as depicted in *2001*, then the next 2 million years were a series of monotonous reruns: The simple, taco-chip-sized bits of chipped rock that *H. habilis* crafted as tools remain unchanged in their design for the million years that the creature roamed the earth. The tools of its descendants, *H. erectus*, were slightly more sophisticated, but these, too, remained equally monotonous in their design for a million years. Even the earliest anatomically modern humans displayed little innovation in their tool-making styles, relying instead on a small group of designs for choppers and cutters that remained unchanged for millennia.

Yet through all this time that our ancestors' tool design was stuck in a technological rut, the human brain was going through the lion's share of its huge expansion. It was only in the so-called creative explosion that our ancestors underwent some 40,000 years ago, which resulted in the first signs of art, music, and culture as we know it, that human technology suddenly blossomed in innovation, with the invention of everything from bows and arrows to fishhooks to metal alloys to jets to personal computers. But ancient fossils reveal that the human brain had already reached its modern size some 60,000 years *before* that happened.

WHAT'S A BIG BRAIN GOOD FOR, ANYWAY?

So, if it wasn't upright walking or toolmaking that led to the expansion of the human brain, what was it? One key to understanding the origins of the ballooning brain is realizing that our species' remarkable intelligence does not make us evolutionary freaks: As a family, primates have been evolving big brains for much of their evolutionary history. On the whole, monkeys and apes perform better on intelligence tests than any other family of animals. If the big human brain is part of an overall evolutionary trend toward intelligence among *all* primates, the real question becomes, What are primates doing with all those smarts that other animals aren't?

The answer to that question lies in how primates—especially human primates—interact with each other. The most important feature of the lives of primates is their social environment: Many species of monkeys and apes live in social groups characterized by alliances, feuds, hierarchical political structures, competition, cooperation, and deception. Primates almost never use tools in the wild, but they readily use each other as "social tools," forming and breaking alliances to achieve a goal. The intricacies of entering into complex relationships with other people placed the largest cognitive burdens on the brains of our ancient ancestors—and fueled its evolution.

Whether it's negotiating a hierarchy among a group of vervets or making one's way through a crowded cocktail party in Manhattan, maintaining complex social interactions makes huge demands on the intellect. Unlike the physical environment that typically surrounds primates, which changes very slowly, a social environment is never constant. Rather, it is a jumble of intrigues and alliance making or breaking among many different players, all of whom are actively engaged in trying to further their own interests. As Lewis Carroll's Alice observed during the croquet game in *Alice in Wonderland,* which featured hedgehogs playing the roles of croquet balls and flamingos as mallets, "You've no idea how confusing it is, all the things being alive."

What's more, social exchanges often involve more than one or two people acting at the same time, vastly increasing the difficulties of trying to figure out the ins and outs of a particular social dilemma. The social complexities involved in whether to team up with the group leader on a project, for instance, was no doubt not much different for our ancient ancestors going on a hunt than it is for people in an office today: Deciding whether to work with your boss on a new project, for instance, involves your considering that your boss will be happy, your coworkers jealous, your spouse angry at the extra time spent at the office, your other clients worried about decreasing attention, your partner unwilling to take up the slack on some of your other projects, and that your boss might take credit for all your hard work; in short, a Byzantine web of considerations and counterconsiderations more complex than any game of chess and as old as humanity itself.

Throughout human evolution, people have had to constantly juggle a wealth of complex cognitive information in their everyday social transactions. Of course, the mental machinery that makes such social interactions possible operates so smoothly that we typically take it for granted. For instance, people routinely measure one relationship against another, as in "John gets along better with his boss than Mary does with hers"; we assess relations according to conventional expectations, as in "Mary is remarkably close to her second cousin Sylvia"; and, most important, we try to understand social relations in terms of goals, desires, and motives: "He just married her for her money"; "She's trying to intimidate you"; or "He is suddenly being affectionate because he wants something."

The human brain is so prone to seeing the world in social terms that even toddlers will interpret a simple action between two inanimate objects as a social interaction involving goals, desires, and beliefs. Studies show that young children watching a video depicting a large ball jumping on top of a smaller ball, for instance, typically interpret the scene as an act of anger by the large ball. Conversely, children will regard a large and small ball gently rubbing against each other as a sign of affection. More important, when children watch a large ball "hit" the small ball, for instance, they expect the small ball to reciprocate by doing something to the large ball. The children's expectations concern the *sentiment* behind the action—whether the ball behaves angrily or lovingly—and not on a particular action such as jumping or hitting. In other words, the children regard a harsh bump by the little ball against the big one as an equivalent retaliation for the big ball jumping on the little one. The human propensity to see the world through a social lens often extends outside the realm of human-to-human relations as well. People routinely yell at their cars, talk to their pets, and curse the weather. People even create social bargains with themselves: promising their hedonistic side a dish of ice cream, for instance, in return for their more Spartan self doing a hard workout at the gym.

THE DEAL-MAKING PRIMATE

Primates are not the only animals to form social relationships, of course. Bee and ant societies are so integrated as social colonies that some biologists regard them overall as a kind of "superorganism." Elephants and dolphins, too, often live in long-lasting social groups. But merely living in a large group is not demanding enough to trigger an explosion in brain size; more important is the kind of interactions that take place once that social structure is in place. Primate groups are not mere collections of individuals, as is the case with schools of fish or herds of cattle. Rather, they are social organizations where individuals engage in a series of complex interactions with other individuals in an ever changing web of social and political intrigue, often spanning many generations.

Primate social groups typically are organized in a hierarchy, and so one of the biggest challenges facing many primates is forming alliances with others who might be beneficial because of their rank or status. "Vervets, macaques, and baboons are—it must be said—dreadful social climbers," Cheney and Seyfarth say. "Vervet monkeys—like characters in a Jane Austen novel—organize their lives around two principles: to maintain close bonds with kin and to establish good relations with the members of high-ranking families." Primates are as status-conscious as any blue

blooded society wag: In one group of monkeys, for instance, a female
dubbed Becky used alliance making, threats, selective grooming, and
playing with the infants of higher-ranking females to gradually move up
the female social ranks. Gradually Becky rose from number forty in the
pecking order to number twenty-one. Becky demonstrated that achieving a higher rank can be done with a simple strategy: *Make friends in
high places, and then throw your weight around.* It is a strategy that should
be familiar to human primates as well, whether they are in a corporate
boardroom, local politics, or a social club.

The research of Cheney and Seyfarth demonstrates that networks of
allegiance—and vengeance—are as much a part of vervet life as they
were for the Hatfields and McCoys. In one episode, a vervet attacked
another individual, and though the incident lasted only a few seconds,
it did not go unnoticed: Twenty minutes later, the *sister* of the monkey
who was attacked in the previous scuffle suddenly attacked the *sister* of
the attacker. In another instance, a scuffle over food resulted in brother
and sister vervets chasing off the offending monkey; at the same time,
the pair's sister attacked the fleeing vervet's sister. These incidents suggest that it is not enough for a monkey merely to learn his or her place
among the social ladder. Each monkey must also be aware of the various
smaller alliances within the group and predict who is likely to come to
someone else's aid.

Primates' social environment is so important that their brains appear
to be "customized" to perform best in social situations. For instance,
playing a recording of a vervet monkey making a warning call that a leopard is nearby results in other monkeys scampering for safety. But when
a stuffed antelope is placed in a nearby tree—a clear sign that the leopard who killed it might be nearby—there is no effect on the vervets' behavior. Likewise, vervets give warning calls to each other to indicate the
presence of a python but seem totally unaware that the large, distinct
tracks made by the huge snakes are also a warning sign that a python is
nearby. "Monkeys have a kind of laser-beam intelligence," say Cheney
and Seyfarth. "While they solve social problems with little difficulty or
training, they often flounder when confronted with the same problems
outside the social domain." The researchers go as far as to suggest that

even if they dramatically changed the vervets' physical surroundings, the monkeys would barely notice. "How would they respond, for example, if we were somehow able to move Kilimanjaro so that it appeared to the north rather than the south, or if we could cause the sun to rise in the west? We expect that the monkeys would remain utterly unfazed."

In contrast, a distinguishing feature of the human mind is our remarkable ability to apply the knowledge gained in one situation to a new problem. Just as a wrench can be used to hammer a nail, a mental organ that has evolved for one purpose sometimes can be used in another domain—even though it may not perform as well at the task. For instance, it is unlikely that the human brain evolved specifically to do complex mathematics—as anyone who has taken algebra may already have surmised. Yet more than any other animal, humans are able to apply knowledge gained in one arena—such as the strategies of medieval warfare—to another problem, such as the strategies of a game of chess. From this perspective, what is most remarkable about how people perform on logic and algebra problems is not how poorly they do, but that, despite having to deal with abstract symbols and rules that have no direct meaning to their lives, they can do it at all.

WEAVING TANGLED WEBS

Having the brainpower to pay keen attention to the details of social interactions is important because, as Niccolò Machiavelli pointed out in *The Prince*, "So simpleminded are men and so controlled by immediate necessities that a prince who deceives always finds men who let themselves be deceived." From telling a white lie about a colleague's new outfit to being suspicious of a phone solicitor's offer that seems too good to be true to trying to find out who among the children broke the new vase, everyday life is a constant spinning and unspinning of tangled webs of deceit and intrigue.

Monkeys and apes appear to be quite adept at deceiving each other as well. In one instance, researchers hidden in the brush watched as a young baboon approached an adult digging for corms, a tuberlike plant prized among baboons and which juveniles rarely get to eat. Even though he had not been provoked or threatened, the young baboon suddenly let out a scream as if he were being attacked, which resulted in his high-ranking mother dashing in and chasing the adult out of sight. After watching both adults disappear into the brush, the juvenile casually sat down to his tasty meal. There are many similar anecdotes testifying to a web of primate duplicity that would have made Machiavelli proud: A female monkey was observed stifling the ecstatic grunts of the low-ranking male she was mating with because the dominant male of the group—who typically frown upon females consorting with other males—was within earshot. In another instance, a female baboon cozied up to a male who had just killed an antelope and, coaxing him to lie down with the promise of sex, suddenly snatched the antelope carcass and ran. In yet another instance, a female baboon took twenty minutes to inch slowly behind a rock where the dominant male could see her head, but not her hands, which were busily grooming a subordinate male who was hidden from view. These anecdotes of Machiavellian deception among primates suggest that one of the most challenging tasks facing the primate brain is both practicing deception and, more important, keeping an eye out for deception by others.

Yet merely deceiving someone does not necessarily require vast mental powers—or even much of a brain at all. An African beetle will cover itself with the carcasses of dead ants to gain entrance into the ant colony for a feeding feast; possums play dead; and a bird called a plover will walk away from its nest pretending to have a crippled wing in order to trick an intruder into chasing it, thereby deflecting the threat from its young. In these instances, however, the act of deception is an innate behavior that is quite limited in how it is used. The plover, for example, will use its deceptive practice to protect its young but not to protect food or anything else. It's a little like people who tell lies only about what they ate for breakfast.

THEORY OF MIND

An important part of the human ability to engage in sophisticated social exchanges—including truly deceiving someone—is what philosophers and cognitive scientists call a *theory of mind:* being able to form a "theory" in one's own mind about what another person is thinking in his or her mind. We're all natural mind readers: People regularly attempt to understand the behavior of others in terms of what they are trying to achieve, what they believe, and their assumptions about what others are thinking.

Being able to predict how one's actions will be interpreted in the mind of someone else is an important skill for navigating the sea of social interactions that characterize human society. An offer to help a coworker who is struggling with a particular project at work, for instance, might be received with gratitude or hostility, depending on whether your coworker believes your intentions are to help his career or display his incompetence to your superior. Likewise, your offer to help comes from your being able to predict the limits of your coworker's knowledge and abilities and compare them with your own. The layers of mind reading back and forth between two people can turn some negotiations into twisted labyrinths. In our struggle to form a theory of mind about others, we try not only to gauge their knowledge but also what they intend to do with that knowledge: Imagine that your auto mechanic tells you that your car will need $800 in repairs; you know nothing about cars, and you also know that your mechanic knows that you know nothing about cars, and so you try to read his mind for clues to whether he might try to take advantage of this knowledge. Is he trustworthy? At the same time, the mechanic, reading your intentions, might take offense at your suspicions—or try harder to hide his duplicity.

Because we are so prone to forming a theory of mind about the intentions of others, a simple conversation between two people can take on more meaning than the actual words used in the exchange, because of each person's knowledge of the motives, beliefs, knowledge, and desires of the other person. Even a seemingly simple "Hello" can denote a significant new development in one's relationship—if it is a greeting from the president of the company who you thought didn't even know you existed, for example, or an icy salutation from a lover, indicating the argument you two had earlier isn't resolved yet.

People take these cognitive miracles more or less for granted, but being able to attribute a particular state of mind must have been a crucial development in the evolution of our ancient ancestors. Since the ability to form theories about another's mind is a powerful tool for maneuvering through the enormous complexities of social exchanges, creatures having such a trait would have had a tremendous adaptive advantage over creatures who could not do so. This is particularly true for the most social species on the planet—ourselves.

BECOMING HUMAN

Like learning a language, the quintessentially human ability to form a theory of mind about another person is a part of any child's development. It typically takes place in three stages: By the age of roughly twenty months, children seem to be aware that they have distinct thoughts, intentions, and desires, and express them verbally. At that age, young children are capable of pretending, such as eating imaginary food, suggesting they can distinguish between reality and what they and others are pretending to believe.

Children's ability to comprehend that others may have thoughts that are different from their own doesn't appear to arise, however, until a child is about four years old. In a classic series of experiments, children were shown a puppet show in which a puppet named Maxi put some candy in a blue cabinet. Then Maxi left the room, and another puppet came in, took the candy out of the blue cupboard and put it into a green cupboard. When asked where Maxi will look for the candy when he comes back, children under four years of age invariably said he would look in the green cupboard, where *they* knew the candy was located. The problem was not a simple lack of memory: When asked, the children said that Maxi had put the candy in the blue cupboard. The experiment suggests that until about age four, children find it hard to distinguish between what they know in their own mind and what they think is in the mind of someone else. Hence the young child who greets her grandmother with "Today we're having a surprise birthday party for you!"

This lack of being able to conjure up the minds of others is apparent when young children try to deceive. To truly deceive someone, a creature first has to understand that someone else can think differently than it does. When asked if she has been in the cookie jar, for instance, a child might try to avoid blame by saying no, but she won't wipe the crumbs off her face. That's because she doesn't recognize that her parents' beliefs can be different from her own, and so doesn't try to get rid of the "evidence." The root of the "age of innocence" is not that young children don't deceive, but more subtly—and more important—that they don't realize that true deception, that is, changing the mind of another person, is psychologically possible.

Research on primates suggests that, like young children, monkeys and apes lack the ability to form a sophisticated theory of mind. In one experiment, for instance, an ape named Sarah was shown a video of her trainers trying to solve various problems—struggling to reach some bananas overhead, for example. Then the tape was stopped in midaction and Sarah was given the chance to choose between three photographs of alternative solutions to the problem, such as the trainer standing on a chair. Sarah often chose the correct "solution" to the problem, suggesting that she understood that the trainer's actions (looking up, reaching, jumping) were the result of his desires and intentions. In other words, she was reading the trainer's mind. Evidence that Sarah was not simply carrying out the steps she would do to solve the problem herself came when the ape was shown a video of the same problem, but with a trainer she didn't like. In this case, she chose the incorrect solution.

This experiment suggests that Sarah attributed certain mental states to the trainer. But another experiment casts doubt about her understanding that the knowledge in another person's mind could be different from her own. In this experiment, Sarah was trained to push a button in her cage to unlock a metal cabinet that lay just outside her cage and in full view. One half of the cabinet was regularly stocked with delicious pastries that Sarah's trainer shared with her. The other half contained a vile assortment of rubber snakes, a cup of feces, and putrid, rotting garbage. The trainer displayed extreme disgust toward these items and handled them with rubber gloves. After being trained to push the button to unlock the cabinet for her daily sessions with the trainer, Sarah then watched as a lab confederate, disguised as a masked "villain," broke into the cabinet while the trainer was out of the room. The villain reversed the items in the cabinet, putting the goodies in the bad side of the cabinet and vice versa. Clearly upset by the intruder's actions, Sarah responded by hostilely hurling toys and other items from her cage at him.

Sarah knew that her trainer disliked dealing with the various offending stuff in the "bad" side of the cabinet; moreover, she also knew that this bad stuff was now sitting in the "good" side of the cabinet. Yet when her trainer came into the room for her regular visit later in the day and headed for the cabinet—and what would surely be an unpleasant surprise—Sarah made no attempt to warn the trainer or attempt to keep her from putting her hand in. Instead, she blithely pushed the button to open the door as she had always done. The results suggest that Sarah did not understand that her trainer had different beliefs about what was in the cabinet than she did. In the same way that children under four years of age do not understand that their knowledge of where the candy is hidden could be different from where the puppet Maxi thought the candy was, Sarah apparently did not understand that it was possible that her trainer's thoughts and beliefs could be different from her own.

A MENTAL NECESSITY

The human ability to develop a theory of mind about others appears to be a specialized feature of the brain that operates independently of other mental activities typically associated with intelligence. Deficiencies in this evolved mental mechanism may be responsible for autism: In one experiment, for example, a group of children with various levels of mental development were tested for their abilities to form a theory of mind: Some children were autistic—a syndrome where children are sometimes highly intelligent but have severely impaired social, communicative, and imagination skills; other children had Down's syndrome, which results in severe mental retardation; and others were average children. The children were shown three different cartoons made up of four panels, each of the panels deliberately scrambled so they were out of sequence. One set of panels was "mechanistic," that is, it showed a cause-and-effect scenario such as a man tripping over a rock. Another cartoon series was "interactive," showing a girl snatching an ice cream cone from a boy and the boy crying. The third set of panels required that the child form a theory of mind: It depicted a girl putting her teddy bear on the ground and then turning to pick a flower. While her back is turned a boy takes the teddy bear away. The last panel in the cartoon sequence shows her staring at the ground with a surprised look on her face. To place these cartoon panels in the proper sequence, children had to understand that the girl expected her teddy bear to be there when she turned around—something the children knew was not the case—and that she would be surprised to find it gone.

When asked to rearrange the cartoon panels into their proper sequence, the autistic children—who were older and ranked higher in intelligence than the other children in the experiments—easily outperformed the other children in arranging the mechanistic and interactive stories into proper order. But they performed worse than all the other children, including those children with Down's syndrome, on the cartoon sequences that required the children to form a theory of mind. Autistic children also performed worse than Down's syndrome children on tests where children are asked where a puppet might look for candy after the object has been moved without the puppet's apparent knowledge. The experiments suggest that the inability of autistic children to understand the minds of others is rooted in a biological abnormality in the brain and may form the fundamental basis of the syndrome. The tragic case of autistic children demonstrates that the ability to form theories about another's mind is an innate part of normal human functioning.

The ability to conjure up the mind of another person not only plays a role in social bargaining but is also a vital part of another quintessentially human experience: teaching. A parent can warn children about the heat of a skillet, even as she is picking up the handle, because she knows the difference between her mental experience and ability and that of her kids. Of course, animals sometimes engage in a form of teaching: A cat will teach its kittens to hunt, for instance, but only through the kittens' observing and mimicking of what she is doing. The ape Washoe, who had been taught sign language, was observed teaching her son to sign. But only in humans is it common that one person observes others, assesses their ability according to some general standard of performance, and actively intervenes to improve that performance if necessary.

Without our ability to form a theory of mind, human culture would not be possible. Much of the world of literature, drama, and humor relies on the supreme ability of humans not only to create theories about each character's mind but also to imagine simultaneously how each of these imaginary minds might view the minds of other characters. The tragic nature of Shakespeare's *Romeo and Juliet,* for instance, comes from a series of misconceptions among the characters that only the audience is aware of. Romeo's suicide is the result of his thinking that Juliet has died, and the audience is aware that if Romeo knew what they knew, this suicide would not have happened. To an audience of monkeys, however, Romeo's actions would make no sense, because they wouldn't be able to distinguish between their own beliefs and his.

THE SECRET SHARER

A part of the very human ability to form theories about others' minds is the ability to assess one's own mind. Philosophers and cognitive scientists are still debating over just what consciousness is—or even whether humans are the only creatures who have it. Unlike self-recognition, where animals are aware that they are distinct from others, consciousness is the more sophisticated cognitive ability of self-*awareness:* A creature not only can recognize that it is distinct from other creatures, it is also able to compare its own mind with the minds of others based on the theories it forms about their beliefs, desires, and knowledge—distinguishing the idea "He is sad," for instance, from the idea "I am sad."

Consciousness allows the comparison of not only two different minds at the same time but also the same mind at different times, enabling us to hold off taking immediate action in order to consider alternatives. Being able to build future scenarios of what might happen as a result of one's actions is a key part of many social exchanges, from deciding whether to accept an offer for a new job to responding to an insult in a tough bar. Glimmers of this quintessentially human ability to build scenarios of the future may have first appeared more than a million years ago in the stone tool and bone sites of a fossil area known as Olduvai Gorge in Tanzania. Mineral analysis reveals that the ancient tool-making hominids who lived there regularly imported rocks from hills that were more than eight miles from their lakeside territory and used them later to butcher meat. On the lush savannas eons ago, these ancestors didn't just invent tools—they invented the *future*.

The ability to be able to build scenarios of the future, form a theory about someone else's mind, and teach others is vital to any social relationship. Indeed, if the human brain could not perform these cognitive miracles, human society could not exist. That's because, unlike other animals, humans often create their most complex social bonds with people to whom they are not related. The intricacies of entering into complex relationships with unrelated people—from choosing the right mate to making a commitment to a fledgling business to risking life and limb to band together to ward off a marauding mastodon—placed the largest cognitive burdens on the brains of our ancient ancestors, and fueled its huge expansion.

More than any other species on Earth, our species' survival in a harsh, dangerous world depends critically on our near total reliance on each other. Faced with a hostile environment, our ancestors banded together to achieve as a group what they could not do alone, just as our communities, businesses, and nations are tied together in networks of mutual dependency today. It is the pervasive human behavior of *cooperation*, not deception, that is the defining characteristic of human existence. The act of cooperating with one another to survive might seem to run contrary to the evolutionary cliches of a world "red in tooth and claw" and dominated by the "survival of the fittest." But a series of ingenious experiments has shown that cooperation may well be the best strategy to follow, even for the most self-centered person in the world. As counterintuitive as it may seem, nice guys really do finish first.

Chapter Three

NICE GUYS

FINISH FIRST

Because I don't trust him, we are friends.

—BERTOLT BRECHT

It was 1915, and Captain J. R. Wilton, a signals officer with the Royal Sussex Regiment of the British Army, was having tea in the dank mud near Armentières, France. World War I had become a trench-lined struggle in which weeks of bloody combat yielded a few yards of barren countryside. Occasionally, however, there were moments of quiet when a soldier could have a bit of tea and relax. Wilton was enjoying one such moment when an artillery shell suddenly screamed overhead and exploded nearby. The British soldiers quickly scrambled into the trenches, readying their weapons and swearing at the Germans. Then, suddenly, "a brave German got on to his parapet," writes Wilton in his diary, "and shouted out '*We are very sorry about that; we hope no one was hurt. It is not our fault, it is that damned Prussian artillery.*'"

Enemy soldiers might seem like the last people on Earth who would cooperate with each other. But, in fact, peace often broke out among the German and English infantry in the trenches of World War I. Sometimes there were truces arranged through formal agreements, but much of the time the enemy soldiers simply stopped shooting at each other—or at least shot where it would do no harm. According to one account by a German soldier, for instance, the English battalion across the trench would fire a round of artillery to the same spot every evening at seven, "so regularly you could set your watch by it. There were even some inquisitive fellows who crawled out a little before seven to watch it burst."

That cooperation would occur among enemy soldiers may have been surprising to the World War I high command, but it isn't surprising to Robert Axelrod. In fact, the results of an extraordinary computer tournament conducted by Axelrod, a professor of political science and public policy at the University of Michigan, suggests that cooperation across the trenches—and throughout the world—is a very natural thing.

Understanding the roots of cooperative behavior is crucial to understanding the evolution of the human psyche, because when it comes to getting along with others, we are the real champions of the animal kingdom. To be sure, cooperation is a characteristic also seen to some extent in our primate cousins: Chimpanzees hunt and forage in groups and many other primates engage in reciprocal grooming. But humans have taken that simple primate behavior and expanded it into something that defines our species. An ant colony or beehive may look like a microcosm of a towering apartment building in Manhattan. But in many ways the apartment building is more miraculous, because the folks inside are linked together with a type of social bond that exists independently of the ties that bind blood relations. The modern metropolitan city would not be possible without the uniquely human social glue that we often refer to with terms such as friends, partners, neighbors, and characterize with words such as trust, civility, decency, and responsibility.

There is nothing unnatural about our amazing ability to be cooperative; nor is there anything about cooperation that runs contrary to the principles of evolution. Axelrod's research demonstrates that being cooperative can be a powerful strategy for getting ahead, even in a world full of nasty people. Axelrod's insights into how such a trait arose during the evolution of our ancient ancestors—and how the trait is carried out by the human brain today as it responds to the world around it—is shedding new light on one of the most challenging questions in the quest to understand the evolution of human behavior.

• • •

FOR THE GOOD OF THE GROUP?

For years many biologists and anthropologists assumed that acts of co-operation and altruism arose for the "good of the species." It was thought that in some species of birds, for instance, parents intentionally keep down the size of their families to ensure that all the birds in the population have enough food to eat. If every bird had as many offspring as possible, went the reasoning, then there would not be enough food for any of them and no one would survive. Thus each bird would have evolved a trait whereby it "sacrifices" its potential offspring for the "good of the species." This reasoning seemed apt: To many researchers there didn't seem to be any benefit that the altruistic trait of "helping others" would confer upon those creatures who did the helping. Indeed, it would seem that the time, effort, and sometimes risk that goes into helping others would put oneself at a distinct disadvantage. But if *everyone* in the group was predisposed to helping each other, went the reasoning, then everyone in the species would be helped and no one would be at a disadvantage.

Biologists now realize, however, that the theory of cooperation and altruism arising for the good of the species is seriously flawed. Experiments have shown that when an extra chick was added to several nests of a species of albatross who typically had only one offspring, for instance, it was the extra bird in that particular nest who suffered, not the population of birds as a whole, because the mother albatross simply could not find enough food to feed her family. More damaging to the idea of cooperation arising for the good of the species are the predictions of evolutionary theory itself. If there were such a thing as a population of birds who limit their offspring for the good of the species, a bird who had some genetic mutation that made it break the rules and have as many chicks as possible would be in a very advantageous position, evolutionarily speaking: It would have more offspring while others were keeping the number of their offspring low. The chicks in that family would have the mutant gene passed on to them, which would likewise produce the same behavior of having as many offspring as possible. Inevitably, the cumulative offspring of these "maximum reproducers" would become a larger and larger proportion of the population, and the "altruistic" creatures—

and the genes for this trait—would disappear altogether. Even if by some miracle a population of totally altruistic birds appeared on Earth tomorrow, evolutionary theory predicts that over time, the altruistic trait would likely be winnowed out of the population.

ALL IN THE FAMILY

Yet denying the existence of altruism for the good of the species still leaves unexplained the fact that everywhere biologists look, organisms are cooperating with each other in what appears to be selfless ways. Bee and ant colonies are full of hundreds of individuals who help their queen reproduce—sacrificing their own reproductive ability in the process. Birds rearing their young receive help from other birds who do not have their own offspring. Deer flash their white tails as a warning to others that a wolf is nearby, risking that in doing so their tail will also attract the wolf's attention. Prairie dogs warn others by barking when a predator is near, drawing attention to themselves.

Biologists explain much of this type of behavior by noting that many creatures who cooperate with each other are genetically related. A deer, for instance, can be considered the embodiment of a particular set of genes, and related deer share many of those same genes—each deer's offspring contains half of the genes of each parent. Therefore it's not so bad, from a gene's perspective, to put one deer in jeopardy if it helps identical genes within a brother, nephew, or daughter to survive and reproduce. In a sense, a deer is simply its gene's way of making more genes, and so the more related one individual is to another, the more likely they will behave altruistically toward each other. As the famed British evolutionary scientist J. B. S. Haldane once joked, "I would gladly lay down my life for two sons or eight cousins."

Animal studies show that this is true: Prairie dogs give more warning barks, and deer are more apt to give warning tail flashes, if they are in the presence of relatives rather than nonrelatives. Bees and ants take this genetic calculation to the limit: The queen is the only one in the colony who reproduces; the workers, all of whom are her daughters, are there to help get a few of their kin—the next queens—into the next generation.

There are other examples of cooperation, however, that cannot be explained by kinship. In the human gut there are millions of cooperative bacteria that help us digest food in exchange for our giving them some of the food to eat. Lichen is a plant-animal coop of fungi and algae, each depending on the other for existence. Crocodiles open their mouths and let a particular species of bird pick bits of food from their teeth—a favor the crocodiles return by eschewing the birds for dinner. These creatures are engaging in what is known as "reciprocal altruism." As opposed to altruism that is directed toward one's relations, this type of "you-scratch-my-back-and-I'll-scratch-yours" relationship exists even among unrelated organisms, because they are mutually beneficial.

Still, the existence of mutual benefits doesn't explain it all: After all, who was the first bird to start hanging around inside a crocodile's mouth? Who was the first crocodile to leave a tasty-looking bird alone? Once the relationship got started, what prevented a hungry croc from occasionally getting an easy meal with a simple close of the jawbone? Most important, how did the most cooperative species on Earth—ourselves—evolve to behave this way in spite of the evolutionary proscription of "survival of the fittest"?

THE COOPERATION DILEMMA

The beginnings of an answer to these questions lie in Axelrod's computer tournament. Axelrod is one of the pioneers of a remarkable new way of studying evolution and human behavior—through the use of computers. In Axelrod's research, the computer becomes the equivalent of a nature park, with computer programs taking the place of living organisms. In this electronic "petri dish," Axelrod can create a population of critters that behave according to certain rules, and have them interact. Because the computer can keep track of all the interactions with lightning-quick speed, Axelrod is able to watch as the populations of "artificial life" wax or wane or disappear. And unlike primate researchers, who can only observe what happens naturally in the wild, Axelrod can play God: He can change the rules by which the electronic animals interact, or he can alter the number of creatures in each population, or he can change the criteria by which the animals live or die. The result is the computer becomes a laboratory whereby various theoretical questions about behavior can be tested.

Axelrod was interested in how cooperative behavior might have evolved, given that in many instances being *un*cooperative can be extremely attractive, evolutionarily speaking—and is often the only "rational" choice. The dilemma of being cooperative in a world of rational and sometimes self-centered people is aptly illustrated by a simple "game" that formed the core of Axelrod's computer tournament. Called the Prisoners' Dilemma, this game is well known among political scientists, economists, and experts in a branch of mathematics and logic known as game theory. The name of the game stems from a hypothetical scenario in which two suspected criminals—Smith and Jones—are in jail as accomplices in an alleged crime. The evidence against the two criminals is not overwhelming, so the district attorney goes to each prisoner separately and offers him a deal: If Smith testifies against Jones, Smith will go free, but his testimony will ensure that Jones receives the full five-year sentence. Likewise, if Jones testifies against Smith, Smith will serve five years while Jones goes free. There is, however, a catch: If each prisoner agrees to testify against the other, the evidence against both will be strong enough that both will be convicted. But since they helped the prosecutor, they will receive a more lenient three-year sentence instead of five.

What should each prisoner do? Each has been told by his lawyer that the evidence against them is so thin that the prosecutor could put them both away for only a year at most, providing neither testifies against the other. To further complicate matters, each prisoner is kept in a separate cell while he awaits trial, so they can't communicate any threats or promises.

It's called the Prisoners' Dilemma because of the paradoxical nature of how each prisoner might work out a strategy for deciding what to do: Smith, for instance, knows that if Jones testifies against him, Smith's best move is to testify against Jones as well. That way Smith will get a reduced, three-year sentence instead of the five years he would receive if he *didn't* testify against Jones. On the other hand, if Jones doesn't testify against him, Smith's best move is still to testify against Jones, because that way Smith would get away without any sentence at all.

Of course, Jones is thinking the same way: No matter what Smith does, Jones comes out better if he testifies against Smith. The result is that each prisoner chooses the "most rational" choice—testifying against each other—and so each receives a three-year sentence. But if they had both kept quiet, they would have gotten off with only a one-year sentence. Paradoxically, by acting in their own self-interest, they both come out worse off than if they had both acted for each other's benefit.

Such "dilemmas" exist everywhere in life, from ants who protect aphids in exchange for nectar to chimpanzees grooming each other to ancient humans going off to hunt to modern-day soldiers squaring off across a no-man's-land. In each case the dilemma takes on a similar structure: Each player has the opportunity to "cooperate" or "defect"— sharing one's food from the hunt, for instance, or saving it all for oneself. The payoff for defecting when the other player is cooperating is high, as is the penalty for being the "sucker" who is defected on while cooperating. But at the same time, the payoff for mutual cooperation is higher than for mutual defection.

To take another example, suppose that Smith is a grocer and Jones needs some food. They strike a deal whereby Jones will give Smith a bag containing $100 cash in exchange for Smith giving Jones a bag containing $100 worth of food. They arrange to meet and exchange sacks, never to meet again. As in the case with the two prisoners, both Smith and Jones have the opportunity to get something for nothing by defecting—that is, by handing over an empty sack. Cool-headed logic would dictate that defecting is the best course of action no matter what the other person does. If Jones is intending to give Smith an empty sack, the best course of action is for Smith to offer Jones an empty sack. And if Jones's sack really does contain money, Smith is still better off, strictly speaking, by exchanging it for a sack that is empty. As in the case of the prisoners, the result is that both Smith and Jones offer empty sacks, and both are ultimately worse off: Smith can't pay his rent, and Jones is hungry.

But now suppose that Jones and Smith are not going to meet one time only, but instead meet once a week indefinitely, exchanging sacks of food for sacks of money. Does the prospect of meeting again and again change the strategies Smith and Jones should follow?

This is what Axelrod set out to explore in his computerized petri dish. Axelrod invited various researchers who were authorities on the Prisoners' Dilemma to go head to head with each other in a computerized "shoot-out." The rules in Axelrod's tournament were similar to that of the food-and-money example: Contestants were asked to create a strategy, embodied in a computer program, that would offer either a "C" for cooperate or a "D" for defect over a series of about 200 exchanges. Each time both programs cooperated they each received three points. If both defected, they each got only one point. If one program cooperated while the other defected, the defecting program got five points and the cooperating program got what is known as the "sucker's payoff"—zero. Each program played against all the other programs, against a clone of itself, and against a program that generated a "C" or "D" randomly. The fourteen "contestants" in Axelrod's computer tournament included programs written by psychologists, political scientists, mathematicians, economists, and sociologists, all of whom had done extensive research on the Prisoners' Dilemma.

Axelrod put the programs into his computerized corral, and when the "dust" had cleared, a clear winner emerged. It was one of the shortest programs in the tournament and used one of the simplest strategies. And that strategy, to Axelrod's and almost everyone else's surprise, was to be cooperative. The winning program, called "TIT FOR TAT," had a devilishly simple strategy: Its first move was to cooperate, then it simply did whatever its partner did on the previous move.

Wondering whether TIT FOR TAT's success was a fluke, Axelrod held a bigger tournament. He wrote up the results of the first tournament, including a discussion of why he thought TIT FOR TAT won and why the other programs didn't. Axelrod invited the contestants from the first tournament to try again and put an advertisement soliciting contestants in several specialist magazines. This time Axelrod received sixty-two entries from six countries.

TIT FOR TAT won again.

• • •

THE RATCHET OF EVOLUTION

To explore how cooperation might have arisen in the natural world, where evolution reigns, Axelrod then put the various programs in a new computer simulation that mimicked the way a population of animals interact and reproduce. The rules of this tournament were similar to those in Axelrod's other contest, except that in this case there were an equal number of all the strategies to start with. Then for each subsequent round, Axelrod made the programs that did well in the previous round "reproduce" in proportion to the number of points they earned; that is, the more points a program scored, the more "offspring" it had in the next round.

As might be expected, programs that defected a lot—one program, called "ALL-D," simply defected all the time—proliferated at first, because they took advantage of cooperative but naive programs. But after a while, these "nasty" programs died out along with their victims.

Once again, when the simulation had run its course, TIT FOR TAT had come out on top. Axelrod found that even if the environment contains nothing but ALL-Ds, a few TIT FOR TATs can invade and eventually take over the whole "colony." And because TIT FOR TAT will always score more points than ALL-D will when playing with another TIT FOR TAT— but never do much worse than ALL-D does playing with itself—once TIT FOR TAT is established in this simple environment, it *stays* established. "It appears," says Axelrod, "that the gears in the evolution of cooperation have a ratchet."

TIT FOR TAT might be a good way to win a computer tournament, but does it work in real life? Axelrod thinks so. Many interactions among animals, people, and nations can be thought of, in a general sense, as a multiround Prisoners' Dilemma. From deciding whether to ask the neighbors over for dinner to entering into a business partnership to nations setting up trade tariffs, there are many instances where one runs the risk of being defected on (the neighbors never ask you to dinner in return or your partner skips town with your money) and many temptations to defect oneself (a high tariff on imports means domestic cars sell better at home). Yet often the people involved in these situations are able to carry on long-lasting, mutually beneficial, cooperative relationships. Axelrod's study gives important clues to the secrets that make cooperative relationships successful. Of course, the hypothetical Prisoners' Dilemma is a bare-bones representation of the complexities of real life; but as Axelrod says, "We could use all the insights we can get."

GAMES WITH ZERO-SUMS

Many such insights come from the TIT FOR TAT strategy itself. What makes TIT FOR TAT a winner? For one thing, it is what Axelrod calls "nice": It never tries to beat its opponent and never defects first.

Of all the principles involved in fostering cooperation, the idea of not trying to do better than the people you are playing with is perhaps the most difficult to grasp, says Axelrod. One reason for this, perhaps, is that most people are used to playing what are called *zero-sum* games. In zero-sum games such as chess and poker, the only way to get ahead is to take something away from your opponent. The less your opponent gets, the more you get. But there are also *non-zero-sum* games as well, where the players still try to get the most points, but often they can do better if other players do better, too. A game like charades, where the players work together to guess a book or movie title, is a little more like a non-zero-sum game.

Axelrod's tournament was also a non-zero-sum game: A program didn't have to win its points by taking them away from another program; instead, points were bestowed on the basis of the contestants' joint behavior. Indeed, because TIT FOR TAT never defected unless defected upon, it *never* outscored any of its partners in their series of one-on-one exchanges. Yet because mutual cooperation pays better than mutual defection, TIT FOR TAT was able to accumulate the highest overall score because it formed cooperative relationships with many other programs, while other contestants defected on one another, thereby accruing fewer points.

While many parlor games are of the zero-sum variety, much of human life is not. An ancient hominid was likely to get more meat over the long run if he shared his bounty with others, who in turn shared with him, than if he hunted and ate alone. Early farmers who cooperated to build irrigation systems no doubt received more payoff for their collective efforts than if they had tried to build the canal by themselves. Coworkers on a big company project may be more likely to get bonuses if the project goes well than if each were working to do the project alone. The mutual benefits of cooperating are so great that laws are passed to force companies in the same line of business to be competitive with each other; otherwise, price-fixing cartels often develop.

A major stumbling block in many negotiations is that the people involved often mistakenly view the deal as a zero-sum game. A buyer trying to reduce the price of a car by $100, for instance, might be viewed by the seller as trying to take $100 out of his pocket. Under these circumstances, each person works under the assumption that the only way to satisfy the other is at his own expense. (A far better approach is for both people to work toward developing an objective way of deciding the fair value of the car.)

Perhaps the biggest obstacle to fostering cooperation is that people sometimes fail to recognize the nature of the game they're playing. Consider the story of an American father who was tossing a Frisbee with his son in London's Hyde Park: It was the early sixties, and few people in England had seen the strange disc in action. A small crowd had gathered to watch. Eventually one of the onlookers came up to the father. "Sorry to bother you," he said. "Been watching you a quarter of an hour. Who's winning?"

NOT-SO-MACHIAVELLIAN INTELLIGENCE

Another reason TIT FOR TAT was so successful was that, contrary to the advice of Machiavelli, it didn't try to be too devious or deceptive. TIT FOR TAT's strategy is transparent, and it quickly becomes clear to most partners what kind of response a particular action will elicit. TIT FOR TAT's strategy couldn't be more obvious: Because it merely mimics its partners' previous actions, playing against TIT FOR TAT is in essence the same as playing against oneself. Other, more complex programs in Axelrod's tournament tried various methods to test their partners' responses, or tried to guess an appropriate time to defect without being punished; but none of these Machiavellian strategies worked very well. One such strategy was so hopelessly complex it finished last—behind a program that simply cooperated and defected at random.

Having a clear, consistent strategy is important in real life, too. People often take revenge on their spouses, for instance, without actually telling them what they were originally angry about. While such behavior may provide some satisfaction, it doesn't do much to prevent the problem from occurring again. Handing out praise or complaints to employees may be a good strategy for a business manager, but if it's done inconsistently the workers may later take a lack of praise or complaint as a sign of dissatisfaction or approval. And having a clear, consistent set of rules and guidelines for a child's behavior is an important part of being an effective parent.

TIT FOR TAT's success is due to the fact that while it may never do better than its partner, it never does much worse, either, says Axelrod. In any single round of Axelrod's computer tournament there is an opportunity for one player to reap a big payoff at the expense of the other player by defecting. But that payoff disappears over the long run, because when the same partners play the game with each other again and again, the "defectee" gets an opportunity for revenge.

Moreover, because TIT FOR TAT immediately answers a defection with a defection, it encourages other programs to be nice—if they can—and reaps the rewards of that cooperation. In a tournament filled with equal numbers of TIT FOR TAT and ALL-D, for instance, a TIT FOR TAT will elicit cooperation with all the other TIT FOR TAT programs while still holding its ground with the "nasty" programs—and win every time. In other words, nice guys really do finish first.

CRUEL TO BE KIND

Yet while TIT FOR TAT may be "nice," it is definitely not a sucker. Behind TIT FOR TAT's genteel, cooperative demeanor is its firm commitment to quickly punish those who are not so nice. "One of my biggest surprises was understanding the value of provocability," says Axelrod, a soft-spoken, unassuming man who is hard to imagine fuming in a wild rage. "I came to this project believing one should be slow to anger. The computer tournament demonstrates that it is actually better to respond quickly to a provocation. If you wait, there is a danger of sending the wrong signal." If an upstairs neighbor suddenly starts practicing his tap dancing lessons at midnight, for example, Axelrod's research suggests that it isn't very effective to simply bury your head in your pillow, hoping that eventually he will stop. If you don't react, not only will he not be aware you are angry but he may take your silence as a sign of approval and encouragement. If you finally do break down his door months later and accost him for keeping you awake all those nights, he may be more confused than apologetic.

But while TIT FOR TAT is quick to anger, its cooperative demeanor eventually shines through, because it is also quick to forgive. As soon as nasty programs show a willingness to begin cooperating again, TIT FOR TAT takes them up on it. Several other "nice" programs in the tournament didn't do as well as TIT FOR TAT because once the other player defected they would "sulk" and hold a grudge, defecting for the rest of the game despite any cooperative overtures by the other side.

The ability to make peace is a crucial component of social relations that is often overlooked. But it is a skill in which humans—and some other primates—excel. In one group of chimpanzees, for instance, two highly ranked male chimps became involved in a big fight, then each went off sulking by himself. Then all of a sudden another chimp in the group who had watched the fracas cried out with a call of excitement, bending down into a patch of bushes. This caused the whole group of chimpanzees to rush over and gather around the excited chimp, including the two males who had been fighting. Then just as suddenly, the chimps slowly began to drift off, leaving the two quarreling males alone in close proximity. They then began to groom each other in an act of apparent reconciliation. Although these chimps were seen going through this scenario several times, the researchers observing the group never saw what it was that caused the chimps to gather round in the first place. In another instance, a female chimp began grooming one of two females who had been fighting. As she was grooming, the other quarreling chimp slowly edged closer, and the groomer began to work on both of them simultaneously. Then one of the quarreling chimps took over the grooming and the "mediator" chimp walked away, leaving the two alone.

These accounts suggest that peacemaking evolved among our ancient ancestors—and some other primates—as a vital component of social relations. Studies of how children develop social skills reveal the importance of learning how to make up after an argument. Researchers examined the social skills of children who grew up in three different kinds of households. In one kind of home, parents rarely if ever fought in front of the children; in another, the parents fought a lot; and in the third, the parents fought in front of the children but, more important, made up in front of them, too. The researchers found that the children who grew up in "reconciling" families were much better at maintaining social relations than the other children—even those who had been reared in households where there were no conflicts at all. Apparently, it is not the absence of conflict per se but the demonstration of how to resolve those conflicts that is the important social skill children need to learn.

INVENTING THE FUTURE

The most important factor in nurturing cooperation is not friendship, trust, and formal agreements but the "shadow of the future." As long as the parties involved know that they will be engaging in similar deals in the indefinite future, cooperation can evolve all by itself. The evolution of cooperative behavior among our ancient ancestors depended on the knowledge that they would be seeing the people they were interacting with day after day.

The same is true in modern society, which despite its large-scale anonymity is full of many subgroups in which people are intimately interconnected. Diamond traders, for example, successfully conduct millions of dollars of business with a verbal agreement and a handshake. Likewise, a study of the records of one manufacturing company showed that fewer than a third of its business agreements were policed by legal contracts. The enforcement for these relations, says Axelrod, comes from the future. Diamond dealers know they will see each other again and again, and so must keep their word if they want to continue their business. According to a businessman in one study: "If something comes up you get the other man on the telephone and deal with the problem. You don't read legalistic contract clauses at each other if you ever want to do business again."

The importance of the future is evident in the fact that, whether in ancient times or the present, as soon as it's clear that the game is about to be over, people's strategy often changes from TIT FOR TAT to "nobody knows you when you're down and out." Studies show that if a manufacturing business gets into financial difficulty, for instance, even its best customers begin refusing payment for merchandise, claiming defects in quality, failure to meet specifications or tardy delivery. Likewise, a study of itinerant workers in California found that while they rarely paid all of their doctor bills, they promptly paid city fines for breaking garbage regulations. Apparently, says Axelrod, the people knew they had to deal with the city on an ongoing basis; since there were plenty of doctors in the city, they were apparently less concerned about paying their bills on time because they could move from one doctor to another.

One way to enlarge the shadow of the future is to increase the number of interactions two parties will have to engage in. The more times the players know they will see each other, the greater the atmosphere for cooperation to develop. Experts recommend spreading out the payment to a contractor for renovating a kitchen over several steps, for instance, so that there is a combination of carrot and stick that rewards the contractor for work completed while retaining the homeowner's ability to withhold money as punishment if something goes wrong. Likewise, the historic peacemaking process between Israel and Egypt took place in small, discrete stages in which Israel incrementally withdrew from the Sinai Peninsula as Egypt normalized relations. Because there was ample opportunity to show good faith—and to retaliate if good faith was broken—the process helped promote cooperation and trust. The World War I trench soldiers' tacit cooperation arose because they faced each other day after day, and so they knew that if they shot at someone today those same people would be shooting at them tomorrow. According to one soldier who kept a diary of the war: "It would be child's play to shell the road behind the enemy's trenches . . . but on the whole there is silence. After all, if you prevent your enemy from drawing his rations, his remedy is simple: he will prevent you from drawing yours." In fact, adopting a strategy designed to damage your partner's chances for survival sometimes can be very risky. According to Axelrod, one of the triggers of the Japanese attack on Pearl Harbor was that economic sanctions, imposed by the United States for Japan's intervention in China, made the Japanese think their power could only diminish in the future. Instead of relinquishing their interests in China, they chose to attack the United States before they became weaker.

IS COOPERATION ALL THAT GREAT?

Axelrod's tournament suggests that cooperation is not just for the chronically nice. The most self-centered individuals in the world, on looking at Axelrod's tournament results, might conclude that using TIT FOR TAT is a good way to further their own interests. The World War I trench soldiers were in a state of war, for instance, but there was a general willingness to, as the infantrymen leaving the front told their replacements, "live and let live." Cooperation can arise among other kinds of adversaries as well: Former major league baseball umpire Ron Luciano, for instance, occasionally would let the catcher call the balls and strikes in games on days that followed particularly epicurean nights. "I'd tell them, 'Look, it's a bad day. You'd better take it for me,' " he writes in his memoirs. " 'If it's a strike, hold your glove in place for an extra second. If it's a ball, throw it right back.' " Cooperation was possible because Luciano, using the unique human power to form a theory of mind, knew that he was in a position to retaliate if he felt he was being taken advantage of—and he knew the catcher knew that, too.

The "selfish" fruits of friendly cooperation are well known to the business world, which is why there are tough antitrust laws to keep companies from conspiring to fix prices or corner a market. The advantages of cooperative business are so great that tacit forms of cooperation between companies arise even without having been explicitly planned out. From oil cartels to cereal makers, the advantages of cooperation are a powerful incentive to get along. It is very hard to distinguish genuine competitive pricing from tacit collusion—precisely because cooperation arises so easily if the conditions are right.

Playing TIT FOR TAT in the modern world poses some challenges that are different from those faced by our ancient ancestors. For most of human evolution, people lived in small groups where everyone knew each other intimately, and the chances were very high that people would interact with each other again and again. In modern society, there are more opportunities to act anonymously and engage with people in one-time interactions. At the same time, however, advances in communication technology and the media make a person's actions potentially more noticeable among more people. Cultural norms have changed, too: The rule of law now makes it possible that individuals are not limited to enforcing social contracts through their own personal resources, because the state, too, now helps punish defectors. But the reliance on law to regulate social exchange has also dimmed the effect of social opprobrium as an enforcer of cooperative behavior. Witness the recent trend among politicians who suggest that "unindicted" is the same thing as "innocent."

THE COOPERATIVE MIND

Starting and maintaining complex social relationships require a good deal of mental sophistication. Consider the number of cooperative bargains you strike simply hosting a dinner party among friends: Extending invitations to your guests involves creating a cooperative bargain based on reciprocity—you invite people who have hosted you on other occasions, or you at least expect that they will invite you in the future. When you buy your food from the grocer, you are engaging in a kind of "fair market" cooperative bargain—you are exchanging resources of equal value. The grocer's faith in your payment with a check is based on past and presumed future transactions, as is your faith in the quality of the food he is offering. And when you finally sit down to the table, you are engaging in a cooperative sharing of food, each according to his or her needs and desires. No one at a dinner party would accuse a guest of violating the bargain by eating too much, nor does the host expect that the guests will offer to pay for their meal.

Despite the enormous complexity of these cooperative relationships, people perform the mental calculations necessary to achieve them with great ease, just as they readily form complex sentences with sophisticated grammatical constructions even though they don't consciously know all the grammatical intricacies of the language they speak. The cognitive work behind such seemingly simple behaviors is enormous, and requires a human-size brain. Ultimately, the real distinction between the behavior of humans and that of other animals is not just that we make tools, but, more important, that we have the brainpower that enables us to have dinner parties.

The ease with which we perform the complex social computations necessary for everyday life should come as no surprise, for it was just these kinds of mental operations that were important to the survival of our ancient ancestors, and so have been carved into the human brain by evolution. Early humans lived in small, tightly knit bands, ensuring that a person who was extended a favor one day—sharing firewood or helping on a hunt—would likely be there to return the favor in the future. Despite the far greater anonymity of the modern world, such close-knit groups still exist. The benefits of cooperation—and potential pitfalls of being a defector—are no less important today than in ancient times, from the sweetheart deals that arise from "old boy" networks to the threat that "I'll see to it that you never work in this business again."

Making these complex social computations requires that the brain have sophisticated mental mechanisms to carry them out. Having such mechanisms is absolutely necessary because, while TIT FOR TAT may superficially resemble a strategy of being cooperative all the time, it is not the same strategy as simply being blindly altruistic. Research with "evolutionary" models of Axelrod's computer tournament demonstrates that a world full of programs that are "naively nice"—that is, they merely cooperate all the time—are quickly pounced upon and wiped out by meaner programs. The key to cooperation, therefore, is not just being nice all the time but also maintaining "eternal vigilance" for those people who are not so nice. That is, one must be on the lookout for both potential cooperators and defectors, a situation that requires far greater brainpower.

One important way people signal their value as a cooperator is by reputation. By behaving cooperatively in the public eye, a person can broadcast the signal that he or she is a good person with whom to form other cooperative relationships. Though behaving in such a way actually might lead to having to forgo immediate benefits in some circumstances, in the long run having a reputation for being an honest, cooperative person will attract other like-minded people who want to interact, eventually leading to higher payoffs. Of course, having such a reputation will also attract those more devious souls who want to play someone for a sucker.

Playing a TIT FOR TAT strategy is an extremely complex task to carry out in real life. Unlike the programs in Axelrod's computer tournament, humans don't always "play" against one player at a time but often must consider the actions of many different people who are interacting simultaneously. People can communicate with each other beforehand, enabling them to make promises, pleadings or threats, and bluffs that may influence their partners' decisions. Also, a real-life payoff scheme is far more complex than the simple awarding of points: The payoffs may not be identical with each interaction, or may not come at the same time, and a payoff considered beneficial by one player might be considered unworthy by another. Lastly, in real life people don't just offer a "C" or "D" as a response, as in Axelrod's computer tournament. It is up to players themselves to determine whether the actions of their partners have been cooperative or not, which can often be a difficult task. These various difficulties make maintaining cooperative relationships in everyday life harder than merely saying to oneself: "OK, now I'll play TIT FOR TAT." The result is that our brains evolved many specialized features that are customized for social interactions.

WHAT'S IN A FACE?

One specialization that helps us maintain social relations is the brain's amazing talent for recognizing and remembering faces. The importance of remembering the results of past interactions in deciding whether to cooperate with someone else puts a premium on being able to remember and identify many different people, and studies have shown that humans have an incredible memory for faces, as opposed to remembering physical objects or numbers, for instance. Experiments demonstrate that people can remember more than 90 percent of the faces with which they are familiar, even if they haven't seen those faces for more than three decades.

Recognizing faces is performed in an area of the brain that operates independently of consciousness. Some people who suffer from certain types of brain damage, for instance, can't consciously recognize people's faces but nevertheless appear to know these people's identities: Shown pictures of loved ones, they can't recognize the face but display marked physiological responses such as raised heartbeats and an increase in skin conductivity. In other cases, patients who have sustained brain damage can recognize the facial identity of a person but cannot connect that identity to the physical presence of that person. Instead, these patients say things such as "the man looks like my husband, but is not him."

Research on the brains of monkeys suggests that the ability to recognize faces is "hard-wired" in the primate brain. Monkeys have specialized nerve circuits in their brain that become extremely excited when the monkey is looking at pictures of monkeys' faces, but not when it is looking at any other objects. Even monkeys who are raised alone, with no interactions with other monkeys, will nevertheless recognize slides of monkeys as being distinct from pictures of other animals. In other words, they have an innate ability to recognize members of their own species even if they've never seen any. After about four months of living in isolation, however, this ability disappears. And when monkeys who are raised in isolation are later put into a group of monkeys, they become social misfits, unable to form relationships with others and engaging in inappropriate and often violent behavior. Thus even though the brain is "wired" by evolution to be able to carry out social skills, it nevertheless must also have social input from the environment for these skills to blossom fully.

Not only is the primate brain particularly adept at recognizing faces, it is also "tuned" to become extremely active when shown other monkey faces making facial expressions of great social significance, such as diverted gazes or yawns—which in monkey society are expressions of antagonism. In the same way, the human mind seems keenly responsive to the slightest facial signal of social significance—such as a wry smile or a raised eyebrow—even if seen from across the room at a crowded cocktail party.

Being able to read the social implications in a face is so important that the mind remembers faces better when they belong to people with whom a person is having a relationship—or even if there is just the potential for having a social interaction. People who are shown pictures of new faces and asked to assess each person's likability or emotional state, for instance, later have better recall of those faces, as compared to when they were merely asked to assign some nonsocial attribute such as age or gender to those faces. In another study, patients with severe amnesia interacted with experimenters who were purposely helpful or unhelpful to them. Later, the amnesiacs couldn't recognize any of the various experimenters with whom they had previously interacted. But when they were shown pictures of the people they had encountered and were asked to judge who looked as if they would be most likely to be helpful, they pointed to those people who had been helpful to them previously. Even though their conscious brain couldn't remember the faces of these different people, part of their unconscious brain in fact did remember the social experiences they had with them as being good or bad.

THE SEARCH FOR COOPERATORS

The ability to "read" a face for signals that indicate helpfulness or hostility has dramatic implications for playing the game of TIT FOR TAT. Unlike Axelrod's tournament, the real world has no requirement that each person play the game with every other person at random. With the ability to identify beforehand which people are more likely to cooperate or defect in an exchange, people can try to interact only with cooperators and avoid potential defectors, increasing the chances for a higher payoff.

To see the effects of being able to select with whom one plays the cooperation game, Axelrod ran another computer tournament that allowed programs to play mostly with one program or another. He found that while a single TIT FOR TAT can't "invade" a whole population of nastier strategies, a handful of TIT FOR TATs that selectively interact with each other can eventually take over the whole population of nastier programs. The key is for the "nice" strategies to recognize each other and interact more often with each other than with the public at large.

Being nice isn't just for sentimental suckers—it is a very effective way of getting ahead. Nor does being cooperative simply mean that you are a pushover all the time. The real challenge of getting into and maintaining cooperative relationships is determining when to be nice and when to be nasty. This, of course, is precisely why the gigantic primate brain arose in the first place: It takes a lot of mental powers to remember the past, weigh the pros and cons of the present, and predict the future. There is another part of our evolved human psychology that lends a hand in making these tough day-to-day cooperative decisions: the emotions.

Chapter Four

———

WHAT'S GOOD

ABOUT

FEELING BAD?

Most of the time we think we're sick, it's all in the mind.

—THOMAS WOLFE

They are mental maladies that, like a pin prick, let us know that we are alive, and very human: a twinge of guilt, a surge of panic, the up-welling of anxiety from being left out of a group, the sadness that accompanies a setback, the euphoria of sudden love. The emotions are the mind's background music, setting the mood and influencing nearly every aspect of our lives.

These days, emotions such as guilt, anxiety, and fear are often regarded as things to overcome—aberrations that have to be fixed with a change of scene or a program of therapy or drugs. But in fact, far from being signs that the mind has temporarily lost its balance, the vicissitudes of our emotional life are perfectly natural, and often useful as well. The various mental states we call the emotions have evolved through the eons to help our ancestors meet the challenges of their times. Just as scientists now realize that physical symptoms such as fever or nausea are useful tools that signal the onset of disease—and sometimes help fight it—the emotions, too, are mental responses that warn of danger, help motivate appropriate behavior in a threatening or welcome situation, and speed recovery from a physical or mental injury. Though the emotions are popularly depicted as mere holdovers from some sort of primeval, animalistic side in our psychology that spring forth and get in the way of our more civilized, rational selves, the emotions are actually part of an incredibly sophisticated social intelligence—one that is most highly developed in humans and our close primate cousins.

Signs of emotional abilities appear in chimpanzees, for instance: They will gingerly care for a disabled relative, and a chimp was once observed rescuing an unrelated chimp from drowning. The ape Koko, who was trained in sign language, once "signed" that she was sad. Chimpanzees also appear to grieve for others. When Flo, one of the apes made famous by the primate studies of Jane Goodall, died, Goodall observed that Flo's young son Flint stopped eating, kept to himself, and sat for hours in the brush, silently rocking back and forth.

Yet for all their seeming ability to feel emotions, apes and monkeys do not appear to be able to make the distinction between the emotions of others and their own emotions. That is, nonhuman primates' inability to form a sophisticated theory of mind about others extends to emotions as well. None of the other apes in Flint's community, for instance, seemed to be at all aware of Flint's dark mood and went about their life as usual—and Flint eventually died of starvation and neglect.

Humans, on the other hand, are intensely sensitive to cues about how others are feeling. Indeed, the emotions that show in someone's face play a crucial role in how we judge a person's goals, intentions, mood, and reliability. Knowing another's mind is key to deciding what kind of cooperative (or uncooperative) relationship you might want to have with that person. Emotions allow you not only to understand the information in what someone says, but also to gauge how seriously the person feels about it—the more emotional the behavior, the more committed that person is to what is being said. So attuned are humans to emotionally charged events that we eagerly seek them out, even if they don't directly relate to us. People gather around fist fights, chase fire trucks, watch soap operas, trade gossip, and read tabloid shock stories about the emotional lives of movie stars they'll never meet.

Emotional cues are so important to human survival that a "universal grammar" has evolved in human facial expressions. The human facial expressions that spring from feelings of grief, sadness, anger, disgust, surprise, fear, and happiness are universal among all human societies. These emotional expressions are hard-wired into the brain: For most people, the facial muscles involved in shaping the face when they are experiencing emotions are not under conscious control. Only 10 percent of us, for instance, can voluntarily pull the corners of the mouth down to make the protypically human "sad" face. The rest of us can make this face only while also moving the muscles near the chin, which is a giveaway for a phony expression. Likewise, only 15 percent of people can voluntarily raise their eyebrows at the center of their forehead to duplicate the forlorn look of grief and distress.

As Charles Darwin pointed out in his 1872 book *The Expression of the Emotions in Man and Animals,* emotional expressions arose not only as signalers of a particular mental state but as signs of changes in a physical state as well. The attack posture of a dog, for instance, with its ears back and fur up, is a result of the dog's physical preparation to respond to a threat. The threat triggers a host of automatic responses in the nervous system, including the release of adrenaline into the blood, a faster heartbeat, and the increased vigilance of the rest of the nervous system. In contrast, the pose a dog strikes when feeling friendly is almost the opposite, with its back down and its tail wagging. In time, argued Darwin, these outward physical manifestations of preparedness to act became signals to others that such an action was imminent. While people sometimes consciously try to hide how they feel about something, these automatic physical responses nevertheless often show through anyway. A person with stage fright might try to appear calm in front of a huge audience, for instance, but the body's physical responses spill over into muscles that are least controlled by the conscious mind, and the person may start trembling uncontrollably.

Precisely because they are hard to control, such automatic emotional responses can signal whether someone is cooperating or defecting. Indeed, knowing that the emotions will be uncontrollably revealed in the face and other "body language" makes someone an attractive candidate for social interaction, because it is easier to read his or her thoughts and feelings and to trust the person's sincerity. Blushing may catch someone in a fib and cause embarrassment, but being known as a blusher might be regarded as a good trait to someone trying to find a trustworthy partner.

The difficulty in faking emotional signals may be responsible for many people's willingness to behave cooperatively even when there is little chance that they will be caught or punished for defecting. After all, it takes a lot of mental work to weave a tangled web of deception: As Mark Twain once quipped, if you always tell the truth, you never have to remember anything. Despite the well-developed human abilities of self-deception and rationalization, it may be very hard to maintain the emotional disposition of an honest person while at the same time behaving like a rogue. To appear honest, it may be very helpful to *be* honest. This is particularly true in a marriage, where the intimacy with which spouses know each other often makes it extremely difficult to hide a guilty conscience or an angry mind.

The power and peril of displaying one's emotions to communicate with others are evident in simple parlor games. Being able to disguise one's emotions—putting on a "poker" face—is no doubt useful in deceiving and bluffing others about one's intentions. But a "poker" face is useful primarily in rare situations such as a zero-sum game like poker, where each person is playing against all the others. In poker, letting one's emotions show could give away the kind of hand that has been dealt and hamper one's play. Most interactions in real life, however, are more like the game of bridge, where players work in cooperative teams. In bridge, for the partners to be able to display their emotions to each other about the content of their hands would be a tremendous help in coming up with a mutual strategy for winning—so much so that displaying emotions is prohibited by the rules of the game. The result is that both poker players and bridge players strive to control their emotions, but for very different reasons: For a "me against them" game like poker, showing no emotion gives an edge against the others in the game; in a cooperative game like bridge, showing emotions is so advantageous that it is prohibited by the rules. The fact that most of life is a cooperative game suggests why the emotions continue to play a major role in our day-to-day affairs.

• • •

SWAYED BY EMOTION

Nowhere is the role played by the emotions more apparent than in the purely self-sacrificing acts that people occasionally perform. History is filled with extraordinary stories of heroism in war or other emergencies and tales of saints and good samaritans. But ordinary people, too, perform "irrational" acts of self-sacrifice every day: Many of us leave tips in restaurants even when we never intend to visit them again, or trudge through a snowstorm to cast a vote in an election—knowing full well that from a purely rational perspective, one vote ultimately plays an infinitesimal role in the final outcome. We also return wallets full of money, give to charities, and otherwise refrain from cheating even when others aren't watching. Ironically, these kinds of everyday behaviors run contrary to the model of human behavior used by economists, who assume that people rationally judge the costs and benefits of their actions.

The key to understanding these "irrational" actions lies in the power of our emotions. We need emotions because, as good as the quintessentially human ability to imagine the future is, it is not perfect. Economic studies have shown that people very often regard a "bird-in-the-hand" as far better than "two-in-the-bush": For instance, when given a choice between immediate gratification and waiting even a brief time such as three days, people tend to take the short route to happiness. Experiments show that people overwhelmingly prefer to take $100 now over $120 three days from now. Evolutionarily speaking, such "discounting" of the future makes sense, because people typically have no way of knowing whether those future payoffs will ever come, and so things that will occur in the present are given much higher value.

This propensity to view interactions in the here and now as more valuable than those in the future can be trouble for forging cooperative deals. When the future seems far away, it has the effect of making even long-term relationships seem like one-time Prisoners' Dilemmas. The result is that there is often a great temptation to take the immediate payoff, even though the consequences of such actions might lead to someone being worse off in the long run. From indulging in that second scoop of ice cream to submitting to an illicit love affair to spending this month's profits instead of reinvesting them, the rewards of acting in the present are far more tangible and therefore harder to forgo for some future benefit.

Emotions help keep these temptations in check by giving us another, more immediate payoff in an interaction whose payoff might lie in the future: how we *feel*. People in social interactions are not guided merely by the abstract calculation of maximizing their gain, they are also driven by the urge to avoid behavior that gives rise to anxiety, guilt, and uncertainty and to choose actions that lead to feeling pleasure, pride, security, and satisfaction. Of course, the kinds of actions that trigger such feelings differ widely, depending on the circumstances—after all, even thieves have an "honor" system. But the overall mechanisms of the mind are the same. The emotions become a kind of mental shorthand for dealing with the enormous complexities of social interactions we encounter throughout life. The general strategy of avoiding painful emotions—guilt and shame—and seeking out pleasurable ones—pride and self-esteem—typically has the overall effect of playing a TIT FOR TAT strategy that produces long-term benefits. Honesty is the best policy, if only from the perspective that it reduces the enormous load on one's mental machinery.

The fact that the emotions are part of the overall equation of cooperative exchanges means that it is somewhat silly to ask the age-old question whether people are "innately" selfish or "innately" altruistic. TIT FOR TAT's power to elicit cooperation derives from the assumption that people act in their own self-interest—and in some cases people *do* consciously, selfishly act in their own self-interest. But acting in one's own self-interest does not necessarily mean behaving selfishly. In fact, as TIT FOR TAT demonstrates, the best way to get ahead is typically to get along.

People are neither innately selfish nor innately altruistic, precisely because following such a simpleminded strategy is a prescription for disaster. Instead, people are guided by specializations in the mind that take in cues from the environment and pursue short-term emotional goals, be it avoiding guilt, getting revenge, pursuing pleasure, or maintaining one's status in the community. What we call human nature is a collection of mental mechanisms, evolved during the times of our ancient ancestors, that interact with the modern-day environment to produce behavior that we sometimes label "selfish" and sometimes as "altruistic." We are not "basically" anything.

Over the long haul, satisfying these proximate, often emotional goals is loosely correlated with following a TIT FOR TAT strategy, but not always. Nor does it have to be, for biology typically favors strategies that are quick, easy, and low-cost ways of producing "pretty good" solutions as opposed to time-consuming, strenuous, and costly methods of finding solutions that are the absolute best. People who leave tips in restaurants they never intend to revisit may be acting "irrationally" in the eyes of a traditional economist, but the costs of such behavior are far outweighed by the benefits of following those same instincts in similar, more important social interactions that have higher stakes and higher payoffs. Likewise, people who risk life and limb to save strangers are following behavioral instincts that, under ordinary circumstances, are far less risky yet extremely rewarding.

BALM FOR THE SOUL

The emotions not only help us negotiate the day-to-day cooperative deal making that characterizes human existence, they can also help provide a warning signal or soothing therapy. Just as pain can send an alarm that the vital functions of the body are in danger, emotional pains such as anxiety or fear are crucial warning signals that one's surroundings are potentially dangerous. One of the most basic of emotional warnings is the "fight or flight" response, which was first scientifically recognized at the turn of the century. The characteristic response of rapid heartbeat, quick breathing, and surge of adrenaline prepares the body for an upcoming escape or battle. One of the strongest testimonies to the intimate connection between the body and mind, this fight or flight reaction needs only an imagined danger to kick into full gear. Anyone who has had a quick scare, followed by an immediate realization that everything is all right, knows the strange sensation of having the physical preparedness for action coursing through the body for a brief moment even though the mind is at ease again.

It is not only the body that goes into high gear when trouble seems imminent: The mind, too, focuses its attention on the situation at hand, ignoring the myriad other troubles, external stimuli, and thoughts that are occupying it at the time. This marshaling of mental resources on a task makes evolutionary sense. Those ancient humans who ignored an oncoming tiger because they were preoccupied with romantic visions of love probably didn't leave many descendants. Of course, those who could not leave thoughts of the tiger behind when the danger was over, so they could concentrate on love, didn't leave many descendants behind, either.

What we fear is a deeply ingrained part of our evolutionary psychology, and we carry the legacy of our ancestors' most worrisome fears with us today. Our evolved psychology primes us to respond to selected dangers in the environment. Lab studies show that it is harder to train monkeys to overcome a fear of naturally occurring dangers such as snakes than it is to train them not to fear an artificial danger such as a gun. Likewise, every parent is frustratingly familiar with the fact that small children will blithely wander into oncoming traffic or stick a fork into an electric socket, while at the same time recoiling from big dogs or spiders and being fearful of going into a dark room. It doesn't take much imagination to realize that large, carnivorous animals, poisonous insects, and dark caves were fears shared by our ancient ancestors, too—and for very good reason. Roadways and electric sockets, on the other hand, have not been around long enough to make an evolutionary impression on our psychology.

Like fear, anxiety plays an important role in influencing our response to stressful situations. As might be expected, one of the biggest sources of anxiety for us today, just as it was long ago, is *other people.* So social is our species that the sources of our anxieties range from not having enough people around to having too many. In our vastly populated, anonymous modern world we tend to forget that during the times of our ancestors, one's family and close associates were absolutely vital for one's survival—and loneliness often meant death, either by starvation or predation. Yet, as the work with TIT FOR TAT demonstrates, being among a large group of people is also terribly taxing on the mind, because of the potential pitfalls and payoffs of cooperative bargaining.

Because the advantages of cooperation within a group are so great, our minds have evolved to be acutely attentive to monitoring which groups of people are forming where—and are also particularly sensitive to signs that we are being left out of these groups. From the "in" groups in grade school and high school, to the clubs, informal networks, and social register of society, to professional associations, business cliques, and presidential kitchen cabinets, all of us no doubt harbor at least a twinge of anxiety that we don't belong in some group or another, as well as feel a smidgen of satisfaction that we have been accepted in some exclusive group somewhere else.

DARWINIAN MEDICINE

The emotions are part of a large group of strongly felt responses, in both mind and body, that evolved to help cure a disease or soothe an injury. In a new field dubbed "Darwinian medicine," researchers are showing that in many cases, modern medical practices can in fact derail these evolved natural defenses and prolong a sickness. The proponents of Darwinian medicine aren't arguing that modern medical practices be overhauled. But they do note that, like the study of every other living thing on Earth, the efforts to understand the human body and disease would benefit from keeping in mind an evolutionary perspective.

Take fever, for instance. For years the standard practice of doctors was to try to stem fevers through aspirin or other drugs—and, indeed, a prolonged, very high fever can be extremely dangerous and result in permanent brain damage. But physicians now realize that low-grade fevers are the body's evolved natural defense against infection. A slightly higher body temperature not only boosts the activity of the body's microbiological defenses that are part of the immune system but also slows down the activity of the invaders. Studies have shown, for instance, that children who have chicken pox and are treated with fever-lowering drugs actually take longer to recover from the disease. An evolutionary perspective also explains why people suffering from infections often experience a drop in iron levels in their blood, making them anemic. Iron is a crucial component for the metabolism of many microbes, so lowered levels of iron in the blood mean that bacteria are less able to thrive. In one study of a group of hunter-gatherers who had chronically low levels of iron because of their sparse diet, researchers found that when the people were given iron supplements to bring their blood levels up, the incidence and severity of malaria and other low-level infections suddenly increased, too.

The body has many other natural defenses against the microscopic invaders that have dogged the human species for eons. Popular depictions of ancient life often show our ancestors locked in a life-or-death battle with a tiger, bear, or mastodon. But, in fact, the number-one predator of the human species has always been bacteria and viruses. While we have lost our fangs and bulky muscles over the eons, the body still has very sophisticated evolved mechanisms against these smaller and more dangerous creatures. Our immune system, of course, is the main army that fights disease. But it has help: Coughing, for instance, helps the body rid the lungs of harmful microbes. Mucus in the nose helps battle viruses that cause colds, and diarrhea can help rid the gut of pathogens that cause illness. These responses are not merely symptoms to be cured, as has long been thought, but the body's evolved defenses against disease.

Our modern environment has created new maladies that were not part of our ancestors' lives. The genes that produce dyslexia, for instance, probably had little impact on our preliterate ancient ancestors. The modern toxins of heavy metals, PCBs, and pesticides may have more effect on our bodies than the many naturally occurring toxins for which we have an evolved resistance. The tragedy of Sudden Infant Death syndrome, or SIDS, may be tied to the modern practice of having infants sleep alone. In hunter-gatherer societies, children typically sleep with their mothers and have a lower incidence of SIDS. New research suggests that SIDS may result in children who lack the ability to rouse themselves to continue breathing during a deep sleep, and the typical tossing and turning of a parent that occurs through the night may serve to disrupt the child's sleep cycle enough to prevent SIDS from happening. The incidence of breast cancer is also much lower among hunter-gatherer societies, where women typically become pregnant soon after they reach puberty and continue to produce children throughout their life. This lifestyle is close to what existed among our ancient ancestors and suggests that the hormonal changes brought about by pregnancy are an evolved defense against breast cancer. This protection is lost in modern industrial society, where women typically conceive later in life and have fewer children.

An evolutionary perspective also helps explain some of the quirky medical complications of pregnancy. While obviously, mother and fetus both have a keen interest in making the pregnancy successful, there are subtle conflicts of interest taking place at a biological level. The mother, for instance, needs to maintain her strength for future pregnancies while the fetus wants access to as many nutrients as possible, without seriously harming the mother. The result is a biochemical tug of war, where both mother and fetus release hormones into their shared biological environment to help achieve their goals. Pregnant women sometimes develop gestational diabetes, for instance, which is the result of a complex chemical struggle gone awry between the fetus and the mother over the level of sugar in the bloodstream. Biochemical signals produced by the developing fetus also act to stop the woman's reproductive cycle, thereby protecting the fetus's own interest. And other hormones released by the fetus make the mother's blood pressure rise, so the fetus has access to more nutrients and oxygen. After the baby is born, the mother feeds it with breast milk—which contains tiny amounts of benzodiazepines, a sedative that is also used in Valium, and which has obvious benefits to the mother.

Even the genes in the fetus itself are in evolutionary conflict. Half the genes of the embryo come from the father, whose evolutionary interests lie in developing the largest, most well-nourished baby possible. But the genes from the mother, while sharing the goal of overall good health for the fetus, also have a conflicting goal of looking out for the mother and her future children. Experiments with mice, for instance, reveal that a particular growth hormone in the developing fetus is produced by only those genes that came from the father—the genes that came from the mother are silent, and do not produce the hormone. Conversely, the mouse has another pair of genes that produces a chemical that helps negate the effects of this same growth hormone—but it is the *mother's* gene that produces this substance, and in this case, the father's gene remains silent.

MATERNAL MENTALITY

The biological interests of mother and child result in subtle changes in the mother's psychology as well. The bouts of nausea and morning sickness that often accompany the early stages of pregnancy may be an evolved defense aimed at protecting a developing embryo. Spicy, bitter, and strong-tasting fruits and vegetables such as garlic, onions, broccoli, and coffee beans typically contain low-level toxic chemicals that help ward off insect and animal pests. These toxins can also affect the fragile development of a tiny embryo during the first fourteen weeks of pregnancy. The result is that over the eons, women have evolved a psychological response that during pregnancy produces a strong dislike for eating foods that contain these substances, thus protecting the fetus. Indeed, studies show that women who experience no pregnancy sickness are two to three times more likely to suffer a miscarriage than women who do get pregnancy sickness. Once this initial stage of fetal development is over, the danger is over, too, and pregnant women's enthusiasm for food returns. This, too, may be an evolved response: After all, once the embryo has become large enough to be out of danger, it needs the mother to eat more food to help nourish it.

A mother's psychological state plays a crucial role in the process of giving birth. Humans are unique among apes in that females often give birth accompanied by one or more friends, relatives, midwives, and other helpers. The impact of feeling the comfort and security of having others close by is ultimately connected to the large human brain size, which not only makes us a highly social creature but also makes the process of delivering a baby even more dangerous for our species. The importance of these social connections is underscored by studies showing that in instances when a women is accompanied by a friend or relative during her delivery, the rate of cesarean sections drops 10 percent and the use of anesthesia drops 75 percent.

EVOLUTIONARY BLUES

The psychological impact of the emotions on biology cuts both ways, of course. Where the presence of a loved one can ease the pain of childbirth, the hostility of an overbearing boss or financial setback can produce a stress response that sends shock waves through the mind and body. Stress has an obvious evolutionary benefit: It throws the body into high gear, helping mobilize the body and mind to get through a crisis. But the aftereffects of having the body in a state of emergency can be harmful. Bouts of stress lead to a depressed immune system, for instance, resulting in a greater susceptibility to disease. Precisely because the stress response allows the body and mind to focus on nothing but an immediate crisis at the expense of the future, long-term stress can also eventually disrupt digestion, cause sleep disorders, and ruin one's sex life.

Unfortunately, our evolved mental mechanisms have left us poorly prepared for the stress of modern everyday life. Our ancient ancestors faced stresses far more menacing than most of us do today, of course. Few of us worry about acute starvation, being overcome by the elements, or attacks by wild beasts. Still, for this most social of social species, the constant stresses of being part of a huge, anonymous society can produce a low-grade, constant stress that our ancestors never knew. Numerous studies have shown that the most important factors in reducing stress include being among loved ones and family, and having the time for peaceful contemplation—a description much more fitting for our ancestors than ourselves. Many of us live hundreds, if not thousands of miles from our families, move from city to city with job changes, leaving our friends behind, and fill our lives with a bombardment of stimuli that makes the notion of contemplation seem archaic. All of this produces an emotional toll that may have its ultimate effects on the body as well as mind. Preliminary evidence suggests a life of emotional stress may lead to an increase in cancers and infections, and may exacerbate the mental deterioration that accompanies diseases such as Alzheimers.

Ironically, another seemingly modern malady, sadness and depression, may have evolved as a response meant to alleviate the pain of stressful times. The typical behaviors associated with sadness—a withdrawal from day-to-day contacts, long, quiet sessions of rumination, a resignation, and a slowness to anger or to change one's immediate circumstances— help prevent a person from spending more resources on an undertaking that is clearly unproductive or dangerous. If you have just been beaten by an adversary, it is unwise to immediately challenge him or her again, particularly if you have been injured or have depleted some vital resource. Depression helps to keep you out of the ring for a while, allowing you to recover and perhaps plan another way to approach a problem.

Depression may also help people get a more realistic perspective on their lives. Psychological studies show that cynics are right: Happy people really are self-deluded. Average, happy people consistently overrate themselves, their talents, and their position in a group. Depressed people, on the other hand, are far more accurate in their assessment of their situation. This is not meant to suggest that happy people are foolish, for it is precisely their ability to be optimistically self-deluded that allows them to get through unproductive troughs that are often part of cooperative bargains. It helps farmers plant again after a failed crop, lovers to reunite for one more try at being together, and hunters and gatherers to return to the valley for the prospect of food. Depression, on the other hand, helps us decide when it is in fact time to give up and move on.

It is this crucial role of sadness in our lives that has led some evolutionary psychologists to ask whether our modern culture's reliance on so-called mood brighteners such as Valium and Prozac are really all that good for us. By taking away the psychological pain that accompanies everyday life, these substances may be depriving us of vital warning signals that all is not right. Evolutionary psychology suggests that anxiety, sadness, and fear played a role in our ancestors' ability to navigate in their social world. By taking away one of the mind's compasses for how well that journey is proceeding, we ultimately may be making things worse.

Of course, people nowadays take aspirin and other painkillers and antinausea and antidiarrhea medicines without apparent side effects. Likewise, millions of Americans use mood-brightening drugs without producing radical changes in their overall behavior. And it may be that the modern world is so different from that of our ancestors, producing anxieties and fears that throw our evolved mental mechanisms out of whack, that such medical technologies are little different in principle from the salt and sugar substitutes we use to overcome the abundance of these substances in modern times. Still, one can't help the lingering thought that the biggest impact of these drugs may be the false notion that "normal" is equivalent to being constantly happy, just as doctors once regarded fever as an aberration to be relieved. Evolutionary psychology suggests that our minds are more complicated than that, because our social lives are more complicated than appears on the surface. The mental ups and downs of life play a crucial role that, while sometimes painful, serve to help us along.

The quintessential human behaviors of getting along with other people, which involve family ties, reciprocal altruism, cooperative exchanges, retaliations, peacemaking, and the emotions, play such an important role in human survival that the brain has been customized by evolution to deal with their complexities. Some of these mechanisms include cheater detectors, our psychologies for doing spatial tasks, our ability to form a theory of mind, and our suite of emotional behaviors that help us be the most cooperative species on the planet. In addition, evolutionary psychologists are beginning to uncover evidence of a host of specialized features of the human psyche that aid in negotiating the sometimes treacherous waters of sex, violence, and other socially important facets of human existence such as language. One group of mental mechanisms is targeted as one of the most important relationships in any human's lifetime, one that is the evolutionary foundation of life itself: finding a mate.

Chapter Five

THE

EVOLUTION

OF LOVE

*I see another law in my members, warring against the law of my mind, and bring-
ing me into captivity to the law of sin which is in my members.*

—ST. PAUL (ROMANS 7:23)

*He kissed me under the Moorish wall and I thought well as well him as another and
then I asked him with my eyes to ask again yes and then he asked me would I yes to
say yes my mountain flower and first I put my arms around him yes and drew him
down to me so he could feel my breasts all perfume yes and his heart was going like
mad and yes I said yes I will yes*

—JAMES JOYCE

"Consider this question," says David Buss. "What would upset or distress you more: imagining your mate having sexual intercourse with
someone else, or imagining your mate forming a deep emotional attachment to someone else?" If you are a man, says Buss, a psychologist at
the University of Michigan, it is likely that you find the idea of your mate
having intercourse with someone else far more distressing—indeed, the
majority of men in a study conducted by Buss did so, too. Conversely,
85 percent of the women who were posed this question in Buss's study
said they found the idea of their mate forming a deep emotional attachment to someone else far more upsetting. It wasn't mere talk, either:
Buss wired up his subjects and monitored their physical reaction to imag-

ining their mates either making love to someone else or forming a deep emotional attachment with another. In the vast majority of cases, men reacted more strongly to imagining their lover having sex with another man, and women reacted more strongly to the idea of their lover developing a strong emotional attachment to another woman. Buss's research gets to the heart of the notorious double standard in Western society, where male sexual infidelity typically is winked at, while female sexual infidelity is often brutally punished. Emotional infidelity, which is more a crime of the mind than of action, is much harder to pin down.

Where do these different attitudes toward sexual infidelity come from? To mainstream social scientists, such differences in the minds of the sexes are due to learning, cultural forces, and socialization. These researchers believe men and women think certain ways because the media, advertisements, parents—somebody other than themselves—influence them to think this way. But to evolutionary psychologists, the propensity of the male and female mind to operate in subtly different ways grows out of the deep, evolutionary legacy left by our ancient ancestors. "A long-standing dogma of this century's social science has been that the nature of humans is that they have no nature," says Buss. But, in fact, while people indeed are the products of their upbringing and their culture, the fundamental differences in people's attitude toward sex can't be fully understood without first focusing on the evolved mental mechanisms that produce the fundamental basis of those attitudes. In the same way, understanding the difference between a taxi and a race car isn't possible without first understanding the basic features of the two cars, such as their engines and suspension.

Buss is a formerly "mainstream" social psychologist who early on recognized the power of viewing human behavior through the lens of evolutionary psychology, and happily crossed over. "Most psychologists tend to prefer modern theories—if it hasn't come out in the last five years it is viewed as antiquated," he says. "But I'm one of the few psychologists who have papers with citations dating back to the 1870s." Those papers, of course, refer to the seminal work of Charles Darwin. Though he has studied mainstream psychological topics such as personality theory, Buss's major contribution to evolutionary psychology concerns the quintessential human talent for deal making applied to the quintessential business of evolutionary existence: sex, love, and mating.

Making love is one of the most important, complex, and perilous cooperative exchanges that any of us engage in during our lives. Loaded with promise and fraught with dangerous pitfalls, love affairs tax our abilities at deal making to the fullest, requiring the complete repertoire of psychological specializations that evolved for cooperation. Our ability to form a theory of mind, for instance, enables us to predict the needs, wants, and reactions of our partner, just as our mental "cheater detectors" keep us on a keen lookout for breaches of faith. The evolutionary rewards—and potential costs—of our sexual relations are so high that love and sex capture a good deal of our everyday mental attention. Small wonder that from gossip to tabloids to epic poetry to romance novels to television soap operas, the particulars of other people's love affairs have held a powerful spell on people's attention for eons. Small wonder, too, that sex and love have spilled over from the realm of mere procreation into politics, power struggles, and other aspects of cooperation that make human social life possible.

Like any cooperative venture, a love affair is a delicate balancing act between a person's needs and wants, producing a tension that has raged since the days of our earliest ancestors. At its heart, human sexuality is and always has been a dialogue of compromise between men and women, a cooperative bargain where neither partner gets everything they wish, yet both ultimately reap far more satisfaction than they could have had alone.

Maintaining a long-term dialogue with a mate is no easy task, because the evolutionary goals of men and women are often quite different—and often in conflict. For decades social science and pop culture alike have erroneously assumed that men's and women's sexuality are fundamentally the same—an infinitely plastic, infinitely expressive sexual potential on which culture, religion, and, as Freud argued, early childhood experiences put an indelible stamp. The stereotypes of brazen men and coy women, the cultural icons of males sowing wild oats and females guarding their chastity, the plethora of cultural messages emphasizing the sexual power of wealthy men and young, beautiful women all have long been thought to be solely the result of "socialization" fueled by advertising and the media.

But, in fact, a host of new research by evolutionary psychologists reveals that our most fundamental sexual behaviors have been shaped by the constant negotiation and renegotiation of the different evolutionary goals and desires of males and females. It is a battle that has raged for millennia, resulting in a delicate, not-so-stable equilibrium of conflict and compromise. The debate over everyday sexual behaviors involving marriage rights, sexual mores, double standards, jealousy, the "seven-year itch," and control over a woman's reproduction may seem to be modern-day phenomena, but all these behaviors have a deep-seated, evolutionary legacy that stretches back to the time of our ancient ancestors.

MALE AND FEMALE MINDS

Just as the sexes are different below the neck, they have different mental "organs" above the neck as well: In the times of our ancestors, men and women faced different problems when dealing with matters sexual. As a result, male and female brains have evolved so that the criteria by which men and women choose their mates, how they respond to infidelity, and what fuels their sexual desires are different, too. Like all evolutionary specializations of the human psyche, the sexual "organs" of the mind are a blend of nurture and nature that are neither rigid, fixed behaviors nor predestined biological mandates. Nor are they merely the manifestations of some sort of sociobiological "drive" of selfish genes as they strive to reproduce themselves by having as many babies as possible. Rather, it is love and libido, not baby making per se, that is the evolutionary legacy handed down by our ancient ancestors. As is the case with most animals today, it was not even necessary that ancient humans were consciously aware of the fact that having sex results in pregnancy. All that was needed was that our ancestors responded to their sexual desires and external cues that their mate was desirable. Those people who didn't do this—spending all of their time and energy perfecting their mastodon stew, say, or spending their time seeking sexual satisfaction with trees—didn't leave any ancestors. The fact that our evolutionary legacy of love, desire, and sex is only loosely connected with pregnancy is aptly demonstrated by the behavior of modern-day couples who use birth control, yet who lose none of their ardor for sex even though they are fully aware that it will not result in "maximizing" their fitness.

It's not enough to merely have a simple strategy such as "have lots of sex with lots of people," for instance, simply because it takes two to tango: Any action by one person is influenced by the reactions of the person's partner, who may have conflicting needs, desires, and goals—and who may react in decidedly un-"fitness maximizing" ways if he or she finds out that you are cheating by having sex with someone else. Nor is it likely that, in a species where a huge amount of investment by both parents is critical to a child's development, merely having lots of babies is the best way to pass one's genes into the next generation. One of the much ballyhooed predictions supposedly arising from sociobiological theory—that males desire many different sex partners because it "maximizes their fitness"—does not reflect "typical," "natural," or "basic" human behavior at all. Such a one-dimensional strategy is far too simple, ignoring myriad complications such as a person's lack of resources to raise such children, the possibility of revenge by one's competitors, the potential for disease, and a host of other factors that make each everyday human behavior a complex interaction of many different goals and strategies.

PICKING A MATE

The different evolutionary strategies in how males and females invest in their offspring are reflected in our modern-day behavior of how we pick our mates. The emotional and physical characteristics that each sex prefers in its mates are a deep-seated part of human psychology, not merely cooked up by the media or Madison Avenue—though advertisers are keenly aware of these psychological preferences and are adroit at exploiting them. Clues to the nature of our evolved sexual psychology were gathered in a massive, cross-cultural survey conducted by Buss. More than 10,000 men and women in over thirty different societies around the world were asked to rank dozens of physical and psychological characteristics they preferred in a mate. The results were the same the world over, suggesting that these preferences reflect a fundamental psychological trait common to the human species as a whole.

The top two preferences chosen by both males and females in Buss's study confirm what most people already probably know from experience: The characteristics judged most important in a mate, whether male or female, are kindness and intelligence. The fact that both sexes would choose these traits as important is not surprising; after all, both are crucial for engaging in the complex cooperative give-and-take that is required in any intimate relationship, especially one that involves the enormous commitment that comes with raising children.

After "kindness" and "intelligence," however, males and females part company in what they look for in mates. In all but one society, for instance, women ranked the attributes of "good earning capacity" and "ambition" in males as more desirable than physical attractiveness; conversely, males considered a woman's youth and physical attractiveness more important than her earning capacity.

These preferences make sense from the point of view of the different strategies men and women employ in their sexual deal making: One fundamental reason for the differences in the behavior of the sexes, outlined years ago by evolutionary biologist Robert Trivers, is the basic biological difference between the male's sperm and a woman's egg. Typically, the large, nutrient-rich egg of the female requires more biological resources to produce: A woman, for instance, typically produces only 400 eggs in her lifetime. Sperm, on the other hand, are little more than snippets of DNA with a tail—a male produces some 300 million sperm in a single ejaculation.

This discrepancy in biological investment between men and women extends to well after conception. While women are limited to having at most about twenty births during their lifetime, the only limiting factor in the number of children a man can potentially have is the number of women he can impregnate. Thus in strict biological terms, women typically have more "investment" per child than a man. Not only is a woman's reproductive capability tied up for nine months during pregnancy but also a human child requires more time and energy to care for than in any other species. Women also are put most at risk by copulation: In the medically sophisticated West, it is sometimes easy to forget that until quite recently one of the leading causes of death among young women was complications in childbirth.

These two very different patterns of biological investment can lead to differences in how males and females approach the mating game. The sex that is required to make the least amount of biological parental investment—in humans, males—can use that extra time trying to have more offspring. Likewise, the sex that puts in the greater amount of investment into each offspring is likely to be more selective when choosing mating partners, because there is more to lose if one chooses a lousy mate. These behaviors are not necessarily the result of maleness or femaleness per se: In some species of seahorses, for instance, it is males, not females, who make the biggest biological investment in rearing offspring. In those seahorses, it is the females who exhibit aggressive courtship behavior, and it is males who are choosy about selecting their mates. This suggests that it is the biological logic of who physically invests more in offspring, not merely gender, that influences a creature's sexual behavior.

These differences in investment suggest that what males and females regard as attractive about a potential mate is far from arbitrary: Because the number of offspring a male can have depends crucially on the fertility of his mate, men would be expected to prefer a mate who shows signs of robust fertility, all else being equal. Therefore, to our ancient ancestors, a woman's attractiveness would be defined by physical signs of youth and health—which was no doubt handy because, back then, knowing a person's exact age was difficult if not impossible. These evolved preferences are still with us today. The pages after pages of advertisements for cosmetics in fashion magazines are no mere quirk of modern marketing: In nearly every society around the world, men regard unblemished, unwrinkled skin, clear eyes, and full lips as signs of beauty in a woman—which not coincidentally are also signs that the woman is young, healthy, and disease-free, and therefore likely to be fertile for a long time.

Men also have a keen eye for a woman's figure: not for whether she is thin or fat, but whether her hips are roughly one third larger than her waist—a proportion indicating a pattern of fat distribution that medical studies have shown is linked to robust fertility. In one study, men choosing from a series of drawings depicting a woman with different hip-to-waist ratios and body fat ranked as most attractive those pictures where the woman displayed this ideal proportion—regardless of whether she was fat or thin. Indeed, even though the Miss America winner has become 30 percent thinner over the years, her hip-to-waist ratio has stayed close to this evolved optimum.

On the other hand, in the times of our ancient ancestors, the attractiveness of a man had less to do with youth and fertility, as men are typically fertile well into middle age and beyond. Because women are physically limited in the number of children they can have, they would be drawn to mates most capable of investing time and resources in their children, all else being equal. For a woman, a man's attractiveness is centered more on his willingness and ability to invest resources, and these are typically indicated by the physical signs that suggest maturity, high social status, and access to crucial resources.

This evolutionary legacy of what we find attractive about a mate persists in our modern-day psychology, even though in the West, much of the evolutionary rationale for these traits has disappeared. Women typically have only a few children, modern medicine has helped extend fertility well into middle age, and as women gain more financial independence, they are less in need of resources from men. Yet the evolutionary legacy of our Stone Age mind continues to fuel our modern-day sexual behavior: What counts most to both sexes, of course, is little different from the days of our ancient ancestors—having the intelligence and kindness to forge the cooperative bonds that are crucial to any long-term relationship. But youth and beauty in women, and status and resources in men, were also of importance to our ancient ancestors, and our modern-day desires for these traits still reverberate throughout cultures around the world.

Some researchers argue that a woman's desire for status and achievement in a man merely reflects women's lower economic status worldwide. But even when women have achieved wealth and power on their own, they nevertheless still desire high status, wealth, and older age in their mates. One study of the mating preferences of women who were in medical school found that even though these women were expecting to have a high amount of resources and financial security, their desire for high-income, high-status males actually increased. Likewise, a woman's desire to mate with an older man appears in a wide variety of cultures. Researchers examined the ages of people from a sampling of marriage records in Seattle, Phoenix, and a remote island in the Philippines. They also looked at singles ads in Washington, D.C., and the Netherlands, and marriage advertisements in India. In every case in every culture, women searched for, or married, men who were slightly older than themselves. Men, on the other hand, generally preferred younger mates, with the age difference widening as the man's age increased. The greatest differences in age between male and female couples were found in the tiny island in the Philippines—far from the reach of Western culture.

The criteria for sexual desirability, while geared toward an evolutionary strategy, are not necessarily fixed in how they are expressed: The psychological rule of thumb—*seek out men of high status*, for instance—is nevertheless still triggered by cultural clues that come from the surrounding environment. Two centuries ago, for instance, well-tanned skin meant that a person labored in the fields; lily-white skin was a sign of social status. By the 1950s, when people had shifted to working indoors in factories and offices, a deep tan suggested a person with a good deal of leisure time, and hence high social status. Now the tables have turned again, as a tanned skin has lost its value as an indicator of social status because of the dangers of skin cancer arising from overexposure to the sun—and thus some people might interpret a well-tanned body as a sign of low intelligence.

The old cliché is that beauty lies in the eyes of the beholder, but in fact people's perception of what is beautiful is more similar than it is different. Around the world, for instance, both males and females prefer mates of average height and weight. A study of how people judge the beauty of faces suggests that a major component of attractiveness is how closely a face matches the average of faces in a community. The faces of 96 males and 96 females, all in their late teens and early twenties, were randomly put into a computer, which merged them into composite photos that represented the "average" of four, eight, sixteen, and thirty-two different faces. These composite photos were then included in an overall sample of faces that was judged by a panel of sixty-five students for attractiveness. The researchers found that the judges nearly always ranked the sixteen- and thirty-two-face composite photos of men and women as more attractive than the individual faces that made up the composites. The research suggests that the human brain has been sculpted by evolution to regard as most attractive those faces that are, ironically, most average. This preference may have evolved in our ancestors because, as representatives of the community at large, these average faces would be easiest to read for subtle expressions of happiness or concern that serve as important social clues. An average face may also signal that a person possessed an average, representative makeup of genes, which would help defend against diseases.

SEXUAL SELECTION

The fact that men and women appear to desire slightly different things in their mate resulted in a different kind of evolutionary pressure on our ancient ancestors. As Darwin first observed in *The Descent of Man, and Selection in Relation to Sex*, the sexual behavior of both sexes creates two kinds of evolutionary pressures that can lead to characteristics that appear to have no particular survival value. These characteristics, which include a peacock's tail, a stag's huge antlers, and the mane of a lion, may be detrimental to survival; for instance, large feathers might attract the attention of predators, or antlers restrict a stag's mobility. Darwin argued that one reason these traits arise is because members of the same sex compete with each other for access to members of the opposite sex. Since in many animals, including humans, one male can impregnate many females, there is always the potential for some males to have a great number of offspring while other males have few or none. There is evolutionary pressure not only to have many mates but also to exclude others from mating. "Thus you get two stags locking horns," says Buss. "The winner gets the female, loser ambles off with low self-esteem— and mateless." Likewise, since females are limited in the number of offspring they can have, there is competition among females for males who seem to be the best candidates to father their children.

The other evolutionary process arises as members of the opposite sex choose a mate with particular traits they find attractive. These traits get passed on to their offspring, and those offspring who have the trait get further selected in the next generation. The two evolutionary processes are linked because the characteristics preferred by one sex wind up driving the competition between members of the other sex. "If it suddenly happened that females preferred to mate with men who walked around on their hands," says Buss, "half the world would be upside down in no time."

The intense competition among both males and females for mates has led to a host of strategies used by men and women to attract a partner, which change depending on whether a man or woman is seeking a short-term or long-term mate. The different elements involved in a long-term or a short-term love affair result in differences in what people consider desirable. For a man seeking a long-term mate, fertility and fidelity are important. For a woman seeking a long-term mate, a man's resources and willingness to commit to a long-term bond are important.

For people seeking short-term mates, however, the things they find desirable are different. Men want to find a woman who is sexually available and willing—though such behavior is abhorred in a potential long-term mate. In short, they want to gain access to as many mates as possible for the least amount of resources. For women, on the other hand, one main goal of a short-term affair is what Buss calls "resource extraction"—the most extreme example of which is prostitution. Studies show that women who are pursuing a short-term strategy say they prefer "big spenders" on a first date. Women may also engage in short-term affairs in order to better gauge the long-term prospects of a partner.

Evolutionary theory helps explain what male rock stars know from experience: Even though a man may display no signs of being interested in a long-term relationship, there will be plenty of women who are following a sexual strategy geared toward short-term results. One reason for this is what is known as the "sexy son" hypothesis. A woman who feels that there are few "investing" men available in her environment might instead attempt to mate with a man who, despite his lack of investment, possesses attractive characteristics that will be passed on to her offspring. Thus the more a man sleeps around, the more attractive he becomes, because the very fact of his success at attracting members of the opposite sex is a trait that a woman may want to pass on to her children.

Another reason that women seek out short-term mates is to bring a better sample of genes into their offspring. Studies show that when seeking a short-term mate, both men and women put attractiveness at a premium. Looking more deeply at people's preferences, however, it turns out that men, in fact, will settle for a lot less pulchritude in their mates than their ideal; women, on the other hand, are choosier, and will not lower their standards for attractiveness in their short-term mates.

Women's quest for good-looking men in their romantic flings is a product of the fact that over the eons, good looks reflected a healthy constitution and a disease-free life. Each of us carries two sets of the same genes—one from each parent. In most cases, the genes are similar and perform virtually the same function. Having two slightly different genes in a set, however, in principle confers a greater resistance to disease: If one gene is knocked out, the other gene can take over. This biological redundancy is reflected in the symmetry in people's faces, which is a large component of a person's attractiveness. The link between good looks and good health persists today: In Buss's study, people who lived in areas of the world that are ravaged by infectious diseases and parasites ranked physical attractiveness higher than those who lived in less disease-ridden parts of the world.

Ultimately, whether individuals pursue a short-term or long-term strategy may depend on their perception of what kind of potential mates are in the local environment. In one study, for instance, women and men were asked to fill out a questionnaire about their general attitudes toward the stability of relationships and trustworthiness of the opposite sex. Then they were asked what kinds of strategies they use to make themselves attractive. The study revealed that those women who in general view the world as full of "dads"—men who show a willingness to invest time and other resources in their mates and their children—will emphasize their chastity and fidelity by dressing modestly and acting deferential. Those women who see the world around them as being full of "cads"—men who invest few long-term resources in a relationship and children—were far more likely to choose strategies such as "I wore sexy clothes" and "I had sex with him."

"Dads," whether from rich or poor backgrounds, were found to be more likely than cads to display their wealth, status, and accomplishments by such things as driving an expensive car and treating their mate at a fancy restaurant. As a group, dads also had the highest grade-point averages—an indication of their striving for success. Cads, on the other hand, appear to be doing other things with their spare time: In the part of the questionnaire on how they advertised their sexual behavior, they overwhelmingly chose the strategy of "I had sex with her," rather than "I let her know I was not promiscuous."

It is likely that most people follow a mixture of short-term and long-term strategies simultaneously. Witness the notorious double standard practiced by both sexes: the man who tries to get a woman to engage in sex with him, then dismisses her as a potential wife because she is unchaste. Likewise, women may desire a husband who is a conservative, staid man who is willing to invest his resources in her and her children, while at the same time harboring a desire for an exciting fling with a handsome, iconoclastic movie star. There is evidence, however, that in some cases, men and women will favor one or the other of these sexual strategies, based on their experiences as children. Studies reveal that girls who are raised in families where there is stress between the parents—a divorce, for instance, or an abusive parent—typically reach puberty earlier and are more promiscuous. It may be that growing up in an environment of marital discord prompts some children to view the world as a short-term mating game.

HOW WE DID IT

Clues to the origins of our modern sexual behavior lie not only in our modern-day exploits but also within our modern-day sexual physiology. As one might expect from a species that could be aptly characterized as the "deal-making" primate, the sexual physiology of men and women reflects a long evolutionary struggle between the sexes. Humans are not merely biological freaks because of their propensity to walk on two legs, make tools, and use language. The way *Homo sapiens* make love is somewhat strange, too. The human penis is among the largest, per body size, in the animal kingdom; the human female has the largest breasts. Unlike many other primates, human females do not have physical displays that indicate when they are fertile, they are capable of mating at any time, and typically mate face-to-face. And whereas only 3 percent of all mammalian species form long-term monogamous bonds between male and female, surveys suggest that more than 90 percent of people the world over marry at least once—though the Western concept of "lifelong marriage" is by no means the norm in all cultures. This long-term bonding is a reflection of another unique quirk of human existence: Human infants are virtually helpless at birth and so require the longest and most intense period of parental care among all organisms.

The origins of these modern-day quirks of sexual physiology lie in the life and times of our ancient ancestors. The skeleton of the human ancestor Lucy suggests that our ancient forebears did not live in a husband-and-wife nuclear family, as has long been supposed. The clue is in Lucy's stature: She was a pipsqueak, standing at roughly three feet tall weighing some seventy pounds, while the males of her species were nearly twice as tall and heavily built. The size difference between males and females suggests that Lucy and her contemporaries lived in social groups like those of modern gorillas, where a group of females resides with an individual male who must defend his mating role against challenges by other males. Because the biggest and strongest male will generally win these challenges and leave the most offspring, over many generations the male physique evolves to be larger and larger. Since females are not engaged in this kind of intense evolutionary competition, their body size stays the same. The result of this two-track evolution is that eventually there is a large difference in body size between the two sexes, as is the case with gorillas and baboons. In those animal species where males and females form lasting pairs, as in gibbons, there is less competition for mates, and body sizes of both sexes are more similar.

Another ancient clue that our ancient ancestors did not have an Ozzie-and-Harriet life-style comes from the fossilized teeth of children. In modern humans the emergence of teeth in children follows a well-established pattern: At six years the first adult molar appears, followed by the eruption of the other adult teeth, from front to back, and by age twenty the last of the wisdom teeth come in. At various stages of their lives children will have a mix of adult and juvenile teeth. Even while a toddler is getting by on its baby teeth alone, its adult teeth are beginning to grow within the jaw. The pattern of tooth eruption in the fossilized jaws of the ancient human ancestor known as *Homo habilis*, who roamed the earth some 2 million years ago, reveals that the children of these hominids developed at a rate nearly twice that of human children today—a development pattern more closely resembling that of modern-day apes—and so were unlikely to require the prolonged care of two parents.

If the physiology of fossils suggests a gorillalike sex life among our most ancient ancestors, the physiology of men and women today suggests that at some time during human evolution sexual behavior changed dramatically. For one thing, males and females are more closely matched in size, reflecting a shift to a more monogamous way of life. Another clue to a change in the way our ancestors made love is the size of the modern human penis. It is far larger, compared to the overall size of a man's body, than the gorilla's. The human male sexual anatomy is much closer in proportion to that of our closest evolutionary cousins, the chimpanzees.

The large sex organ of the human male may be the result of our ancient male ancestors switching their sexual strategy from making war, as it were, to making love. When one of the females in a group of chimpanzees becomes fertile as part of her regular cycle, she typically mates with nearly all the other males in the group, who often "line up" for the privilege. Though the most dominant chimp tries to monopolize the female during the very height of her fertility, another form of sexual competition nevertheless takes place—not among the males themselves but among their *sperm*. Capable of surviving several days, the sperm of each male competes within the female for the prize of fertilizing the egg. Indeed, only a very small proportion of the sperm ever make it to the egg; the rest are apparently there to serve as "blockers"—also known as "kamikaze" sperm— that slow the progress of other sperm that might be swimming up from behind. In such a competition, having a head start—and lots of "contestants"—is likely to be helpful. Hence the evolutionary pressure for a larger and larger penis and testicles.

The large penis of the modern human male suggests that at some point our species went through a period when a similar type of "sperm competition" took place—reflecting a shift from a single male–multiple female type of social structure to that of multiple male–multiple female social grouping found in modern chimpanzees. This evolutionary legacy is reflected in other aspects of the male sexual anatomy as well. The male sexual psychology has evolved to respond to the prospect that his mate has been inseminated by another man during his absence. For example, the volume of sperm that is delivered in a man's ejaculation is nearly three times higher when he and his wife have been separated for a long period of time—time during which she may have been impregnated by someone else. The amount of sperm a man ejaculates while having sex with his wife upon her return is unrelated to how long it has been since he last had an orgasm; the important factor is how long it has been since he last had sex *with his wife.* In another study, men were interviewed about their feelings of security concerning their wives' fidelity. When these men's sperm count was measured, the researchers found that the more insecure a man was about his wife's allegiances, the higher his sperm count, again suggesting that the male mind has evolved to respond to the possibility of competition from the sperm of other males.

A woman's modern-day sexual psychology reflects an ancient legacy of sperm competition as well: In one confidential study that tracked the sexual activity and menstrual cycles of nearly 2,000 women who said that they had steady lovers, researchers found that the women displayed no pattern in when they made love to their steady partners. But those women who had outside affairs did so near the height of their fertility, even though they almost certainly did not consciously plan this. Nearly 10 percent of these elicit trysts occurred within five days of the woman having made love to her steady mate—time enough for the sperm of both men to be in competition.

These features of modern male and female mating psychology suggest that at some point in the evolutionary history of the human species, a female was likely to be inseminated by many different men during the same brief period of time. In other words, males and females switched from living in societies similar to gorillas to a social structure that was more similar to that of modern-day chimpanzees. One clue to when this switchover might have occurred comes from the fossil skeletons of the successor to *Homo habilis,* who is known as *Homo erectus.* In this ancient ancestor, who walked the earth from about 1.6 million to 400,000 years ago, males and females were much more similar in height. This suggests that the evolutionary pressure of competition among males for mates—and difference in male and female body size and the gorillalike social structure that went with it—may have subsided by the time of *Homo erectus.* This lack of direct competition for mates resulted in male and female body sizes being more similar, even as the pressure of sperm competition led to the evolution of different male and female sexual psychologies.

All this might seem to signify an evolutionary legacy that results in our desires to have wild orgies of group sex—which is in fact part of the human sexual repertoire. It is also true that both men and women can be sexually aroused by watching others engage in sex. The modest and sedate sexual behavior typically exhibited by most people today, however, suggests that a further evolutionary wrinkle was put into our sexual behavior somewhere along the human saga. That wrinkle was spawned by our unique abilities of cooperation.

Humans, chimps, and gorillas are unusual among the primate family in that it is females who leave the group they were born into when they reach sexual maturity. The emigration of males or females out of their family group serves to prevent inbreeding among closely related individuals, because breeding with close relatives raises the risk of genetic diseases. In most primate species, adolescent males must leave the group they were born into and emigrate to another group upon reaching maturity. But in gorillas, chimpanzees, and humans—the most closely related primates, evolutionarily speaking—it is the females who leave their home group. Thus the males in the home group form lifelong, tightly knit coalitions.

The shift in body sizes in *Homo erectus* suggests that the males' more

egalitarian sexual strategy may have arisen along with a shift in their behavior to include more cooperation with each other in many areas of their
lives. There would have been an increasing need for cooperation for mutual defense from attacks by other groups—as is the case with modern
chimpanzees—and to coordinate hunting. As the needs among males to
cooperate with each other increased, however, it put men in a conflicting situation: Men who engaged in open, direct competition with each
other for sexual partners would run the risk of souring the cooperative
bonds that were necessary for their survival in other arenas. Thus women
became potentially dangerous to men's cooperative bonds—an evolutionary legacy that forms the basis for the modern-day "all male" club.

As male-male coalitions became more important to their survival, they
shifted into an even more egalitarian sexual strategy—egalitarian for
men, that is, but not necessarily for women. Males' solution was to grant
each other "mating rights" over females, giving up constant competition
for all women in exchange for monopolized rights to one woman, as in
the modern-day institution of marriage. This male strategy roughly fits
in with females' sexual strategy as well, though for very different reasons.
And as the plight of women the world over attests, such mating rights do
not always work to women's benefit. Nor do men by any means necessarily practice what they preach: It may be in a man's best interest to
publicly denounce adultery, his private behavior may be quite different.

HIDDEN SIGNALS

Women, of course, have never been docile, passive players in this evolutionary game of sexual strategizing, and this is reflected in their sexual physiology, too. Unlike many mammals, women do not signal the
various times of the year when they are fertile. In some baboons, for instance, the females' genitals swell and change colors. Other animals, such
as dogs, give off distinct odor cues that they are in "heat." Women, however, send out no such signals, and so in theory are capable of mating at
any time.

A woman's hidden fertility may have arisen as part of the subtle power
ploys that have characterized the deal making between males and females throughout prehistory. Since women make the biggest investment

in children, they want to ensure that they get help from the father in raising a child; at the same time, men want to ensure that if they do help raise a child, that child is theirs and not the offspring of someone else. From the male's perspective, therefore, it would be better if a woman underwent a pronounced, periodic signaling that she is fertile, because it makes it easier for him to ensure paternity. He can ward off advances of other males during the short time period she is fertile—and desert her the rest of the time. In other words, if a woman displayed overt signs that she is fertile, a male would be motivated to stay by her only during the short time each month that she could conceive.

The fact that female fertility signaling might be good from a man's perspective doesn't mean that it is best for the female. From a woman's evolutionary perspective, having a pronounced signal that she is fertile makes it less likely that a male will stick around to help during those times when a female is not fertile—which could make life difficult when she is pregnant, for instance. By hiding any overt signals that she is fertile—or falsely signaling that she is fertile all the time, which has the same effect—a female keeps the male guessing as to when she can conceive. This makes it more likely that the male will stay around her more of the time. Needless to say, a woman's hidden fertility also gives her an edge in another aspect of the deal making logic of sexual relations: Hidden signals make it hard to know for sure when a child is conceived, enabling a woman to become impregnated by another male without her partner's knowledge.

THE FEMININE MYSTIQUE

Such shifts in the balance of power between men and women in their quest to gain an edge in their sexual deal making may be what is behind two other mysteries about female sexuality: breasts and orgasms. Human females have larger breasts than any other primate; in most other species, the breasts swell only when the woman is pregnant or lactating. In humans the size of the breasts has no biological relation to the amount of milk they can produce—in fact, large breasts make it harder for infants to suckle. In the past, researchers have suggested that large breasts served as indications to potential mates that a woman had large amounts of stored fat, which would have been useful in ancient times. Others suggest that large breasts were preferred by our ancestors because men mistakenly associated large breasts and nurturing abilities. In his book *The Naked Ape*, Desmond Morris suggested that a female's large breasts mimicked the buttocks, which propelled men to switch from the front-to-rear posture most primates use for mating and induced them to mate face-to-face, which helped to develop the male-female "pair bond." Primatologists are quick to point out, however, that gibbons, too, mate for life, and female gibbons do not have large breasts, nor do they mate face-to-face, and they copulate only when the female is in estrous. And members of a species of chimpanzee known as bonobos mate face-to-face and copulate frequently, but do not form long-lasting pair bonds or have large breasts. It is possible that large breasts would have at first been a turn-off to males because they signaled that the female was pregnant or lactating, and so would not be a good candidate for mating.

It is more likely that the evolution of the big bust came about as part of a female strategy to help keep males honest in their sexual bargains. Since swollen breasts are a signal of pregnancy, they would have served as a signal to a male that a woman was no longer fertile, and so the man would be free to leave the woman unguarded and go off in search of other mates, leaving the woman to fend for herself. By having perennially enlarged breasts, women continually signal that they are pregnant even when they are not, and so the signal loses its value as a clue for men. The result is that men stick around to keep up their end of the reproductive bargain by helping the female.

Of course, eventually a woman's pregnancy would be noticeable, large breasts or not. But research on modern primates suggests that a woman's ability to conceal a pregnancy, even if only for a few months, is a helpful strategy in her sexual deal making. Males typically are reluctant to help raise children they did not father—in lions and some primates, for instance, a male who takes over a group of females will kill all the offspring sired by a previous male. In some species of apes and monkeys, pregnant females attempt to get around this strategy by undergoing a "false estrous," suddenly displaying outward signs of fertility if a new male suddenly becomes dominant in the group. This bit of biological deception allows the female to mate with the newly dominant male, making him believe that her offspring is his, and thereby protecting it. Likewise, having a few months when a pregnancy is concealed may have served a similar function in our ancient ancestors, allowing a pregnant woman to mate with another male, increasing the chances that he will think the child is his so he will help raise the child, and reducing the risk of infanticide.

THE BIG O

A woman's desire to have ultimate control over her reproductive abilities lies behind the difference in men's and women's orgasms. For years Western scientists thought a woman's orgasm did not exist, then Sigmund Freud muddied the picture by arguing that women have two kinds of orgasm, clitoral and vaginal, and that as they mature they shift from the former to the latter. Then came the famed sex researchers Masters and Johnson, whose research revealed that not only was a woman's orgasm centered on her clitoris, but also that some women's capacity for orgasm far exceeds that of men.

In the past, it has been argued that a woman's orgasm need not have any "evolutionary" function at all: A woman's clitoris (and the orgasm it produces) may be no different from a man's nipples: They are both evolutionary side effects of a biological process by which a generic human fetus differentiates into male or female in the womb. Just as a woman's functional breasts become a man's vestigial nipples, the area of the fetus that gives rise to the penis in the male becomes the clitoris in the female, an organ that, while undeniably providing pleasure, need not have arisen to serve a particular evolutionary function. A woman's clitoris may produce sensations that lead to orgasm, but it may not have been "designed" by evolution to do so.

New research reveals, however, that in fact, whether—and when—a woman experiences an orgasm has a big effect on the fate of a man's sperm inside her reproductive tract. During any bout of love, a certain portion of a man's seed will be ultimately ejected by the female. Studying the volume of this effluvium in a sample of volunteer couples, researchers found that if a woman experienced orgasm within one minute before the man ejaculated—or within sixty minutes afterward, the amount of sperm that was rejected dramatically dropped. The findings reveal that a woman's sexual pleasure does indeed have an evolutionary function: controlling the fate of her mate's sperm.

Of course, neither men nor women consciously planned to evolve the sexual behaviors that characterize our everyday lives. All that was required was that those infants whose fathers stayed around their mothers eventually produced more evolutionarily "fit" offspring, leading to the evolution of the stick-around father. At the same time, those females who continually advertised that they were fertile and hid their pregnancies by having perennially enlarged breasts were more likely to have their mates stay nearby. The most important requirement for such a scenario—often overlooked in the heated debate about the origins of human sexuality—is that such a strategy would not have been effective if human infants, and by extension their mothers and fathers, did not directly benefit evolutionarily from both parents cooperating to help raise their children. While help no doubt also came from a child's extended family, the father played a role, too. It is possible for a man or woman to raise a child as a single parent, especially in the modern world, where a person has access to more resources and child care. Evolution merely creates the mental mechanisms that make cooperative bargaining possible—there is no "gene" that makes men and women live in an Ozzie-and-Harriet type family structure, as modern-day kibbutzes and single parents demonstrate.

Ultimately, our sexual relations can be characterized as an evolutionary "Prisoners' Dilemma," where each sex cooperates—and compromises—to produce something they could not achieve by themselves: Males get better assurances of paternity, females get more help, and their children become more fit than children raised without two parents. Seeing the evolution of sexual behavior in terms of forging a cooperative relationship highlights the fact that cooperation need not necessarily be nice, altruistic, or naive. In sex, as in other relationships in life such as those in business, neighborhoods, or governments, cooperation is often the best choice among many competing strategies, some of which might at first appear to offer attractive payoffs but in reality seldom do.

WHAT WOMEN REALLY WANT

As women have become more economically independent in modern societies, they are returning to the kind of relationship with men that our ancient ancestors engaged in for much of human existence—one that is based on a more egalitarian, TIT-FOR-TAT cooperative bargaining. Ironically, evidence for this change lies in the rising divorce rates in modern industrial societies. In hunter-gatherer societies, for instance, couples typically join together for only about four or five years—which is long enough to wean a child. After that time, men and women are more likely to leave each other to find a new mate, if they so desire. Our ancient ancestors, who lived as hunter-gatherers for most of human existence, may have had the same kind of pattern in their couplings. Helen Fisher, an anthropologist at the American Museum of Natural History, suggests that the famed "seven-year itch" in modern society may be a vestige of this ancient pattern of serial monogamy among our ancestors.

With the beginning of agriculture, some 10,000 years ago, however, the choices of men and women suddenly changed. The sedentary lifestyle that came with agriculture meant that men and women could no longer simply split up and go their separate ways if they had differences, because both their fates were intimately tied to a particular piece of land. Hence, men and women were economically tied to each other, and so breaking up would be disastrous. The current rise in divorce rates in Western society may reflect the growing economic independence of women, who are no longer by necessity tied to a single relationship, and thereby are exercising more control over who they mate with, and when.

CONTROLLING INTERESTS

Control over her reproductive fate was a driving force in producing a woman's evolved psychological mechanisms. This evolutionary legacy continues to affect male and female behavior today. While studies show that both women and men can become sexually aroused by watching explicit films of couples making love, for instance, males are more aroused than women by photographs that merely display the nude body of a member of the opposite sex (in the case of *Playgirl* magazine, which contains pictures of nude men, the readers often are gay men). The reason for this discrepancy in the male and female responses to nudity is that for our male ancestors, sex with strangers was a low-cost way to further their reproductive success, and therefore evolution favored men who were aroused by anonymous sexuality. For women, on the other hand, sex with strangers is riskier, because bearing a child sired by an unfit or unhelpful male is very taxing for a woman. While women can be sexually aroused by pornographic movies and romance novels that depict couples making love, they are less excited by the image of anonymous nude males, because to have such a trait would be risky, evolutionarily speaking. After all, for an unfamiliar woman to display her naked body to a man is typically interpreted by the man as a sexual invitation. But for an unfamiliar man to display his genitals to a woman is typically interpreted by the woman as a threat—because the woman's control over her reproductive abilities is put at risk.

The apparent difference between the two sexes' desires for anonymous sexual encounters is known as the "Coolidge effect," named after a famous, probably apocryphal, story about the President and his wife. According to the tale, the Coolidges were touring a farm when Mrs. Coolidge was shown a rooster who was busily mating with a hen. Told that the rooster performed this act as much as twelve times a day, Mrs. Coolidge reportedly replied, "Please tell that to Mr. Coolidge." Later, as President Coolidge was observing the same rooster, he was told of the bird's sexual prowess. "Same hen every time?" the President reportedly asked. Told that it was a different hen each time, the President replied, "Tell *that* to Mrs. Coolidge." Males' greater desire for anonymous sex was revealed in a study by Don Symons, an evolutionary psychologist who has long studied the evolution of human sexuality. Both men and women were asked: "If you had the opportunity to copulate with an anonymous member of the opposite sex who was as physically attractive as your spouse but no more so, and as competent a lover as your spouse but no more so, and there was no risk of discovery, disease, or pregnancy, and no chance of forming a more durable liaison, and the copulation was a substitute for an act of marital intercourse, not an addition, would you do it?" Symons found that men who were engaged in a steady relationship were four times more likely than women with steady mates to answer "certainly would." Among men and women who were unattached, men were six times more likely than women to answer "certainly would," and more than twice as many women answered "certainly not."

The difference in interest in anonymous sex is also reflected in men's and women's sexual fantasies. "Sexual fantasy provides a window into a private world that is potentially unconstrained by real world considerations," says Symons. "In sexual fantasy you can have anything you want, do anything you want, and be anything you want." One study found that nearly a third of the men questioned said that during the course of their lifetime they had fantasized about sexual encounters with more than a thousand different partners, while less than 10 percent of women said this. Nearly three-quarters of the women said they typically fantasized about having sex with someone they loved, or would like to become romantically involved with, whereas men were split nearly equally among fantasies about lovers, potential lovers, and anonymous mates. Women's fantasies were more likely to emphasize touching and the personality traits of their lover, and involve a longer emotional buildup. Men, on the other hand, typically focused more on visual imagery of sex organs, moved quickly to explicit sexual acts, and often involved partners with whom they merely wanted to have sex.

These differences in fantasy are reflected in erotic films and literature: In erotica geared toward men, women are typically depicted as lusty, aggressive, and enjoying sex for sex's sake, without emotional attachments and courtship. In erotica geared toward women—that is, romance novels—sex is part of a greater theme of love, where a man is consumed emotionally by passion for the heroine and no one else. For the heroine, sex is not an act of submission but an act of control, as she masters her man's emotional fate. In other words, erotic materials for both men and women typically present the opposite sex as a caricature of the consumer's own sexuality.

One intriguing reflection of the different attitudes of men and women toward having sex with strangers is homosexual relationships, in which males and females are more free to express themselves unfettered by the conflicting evolutionary strategies of the opposite sex. Studies have shown that homosexual males typically have very many sexual partners during the course of a year—though the prospect of AIDS has changed behaviors somewhat. In addition, many gay male relationships begin with the mutual desire for sex, and that sexual relationships between gay men may be exceedingly brief. In contrast, lesbians, while just as sexually active as men, more often form long-term relationships and are less likely to seek sexual satisfaction with strangers.

The way men and women behave in homosexual relationships suggests that, in heterosexual relationships, a man's and woman's sexual behavior is a compromise between their different, sometimes conflicting sexual strategies, as each poses limits on the other. "Heterosexual men would be as likely as homosexual men to have sex most often with strangers, to participate in anonymous orgies in public baths, and to stop off in public restrooms for five minutes of fellatio on the way home from work," says Symons, "if women were interested in these activities." The fact that women do not typically engage in such activities by no means implies that their sexual drives are somehow less strong than that of men—only that they desire more control over whom they have sex with.

WOMEN AS PROPERTY

Just as men's sexual behavior is curbed by the conflicting interests of women, so, too, is a woman's sexuality confined by the wishes of men. Women have a strong desire to control whom they mate with because they make a comparatively larger reproductive investment in children. But men, who have much to lose if their mate bears the offspring of someone else, will also strive mightily to control the reproductive choices of women.

The battle for control of a woman's reproductive ability is as fresh today as it was on the ancient savanna: The impassioned debates that rage today over sexual harassment, women's financial independence, and abortion have at their roots the long evolutionary legacy of both men and women striving to negotiate, subvert, and control the reproductive careers of women. These issues are not merely the result of an innate, arbitrary "evil" or obtuseness on the part of one sex or another. As the complexities of cooperative relationships suggest, they are the consequence of an ongoing conflict over the most important facet of both men's and women's lives, evolutionarily speaking.

One of the clearest, most troubling examples of evolutionary mechanisms running headlong into modern society is in how men and women act when they are jealous. The evolutionary logic for jealousy is straightforward: A man who puts substantial resources into a child sired by another man experiences a drastic loss, evolutionarily speaking, because he is putting effort into helping someone else's offspring. A woman, on the other hand, is most affected by the loss of her mate's help and resources, and so can be expected to respond negatively to clues that her partner is sharing resources with someone other than herself. The result is that men are more concerned that their mate has had sex with someone else, while women are typically more concerned with how their mates dispense their resources and attentions. In other words, a woman has the most to lose if a man abandons her for someone else, and men stand the most to lose, evolutionarily speaking, if their partner gets pregnant by someone else and stays. As Samuel Johnson put it, the difference between male and female infidelity is that "the man imposes no bastards upon his wife." These views on what triggers sexual jealousy in either sex have been confirmed by psychological research: As Buss's studies have shown, men are more distressed over imagining their lover having sex with another man, and women are more upset about the idea of their lover developing a strong emotional attachment to another woman.

The result of men's intense preoccupation with their mates' having sex with others is that throughout history and around the world, men have labored to control women's sexual behavior—through chaperones, chastity belts, veils, guarded harems, strict laws against adultery, and by outright threats of violence in retaliation for infidelity. Moreover, in dozens of societies around the world, tens of millions of women are subjected to ritual mutilation to their genitals in a brutal attempt to discourage sexual adventures. Even in the few cultures that have been characterized by anthropologists as being so libertarian that they are free of jealousy, such jealous conflict in fact does exist. Reexamining the anthropological studies of these supposedly jealous-free societies, evolutionary psychologists Martin Daly and Margo Wilson found that while a society may have no customs that punish adultery, for instance, a man will nevertheless refuse to support a child if he believes that it is not his.

The result of men's overriding concern about paternity is that in many societies, women are regarded almost as if they were property. The most exaggerated example of this is in those societies where despots have kept harems of women not merely for reproductive gain, but as commodities in themselves—status symbols in a male's competition with other males. Harems reveal that the ultimate goal of men's sexual behavior "is not simply limitless sex, nor even limitless variety of partners," argue Daly and Wilson, "but the sexual *monopolization* of women, and the more the better."

As much as men may wish them to be, however, women are far from passive role-players in sexual relationships. Efforts by men to control their mate's reproductive career are particularly difficult to achieve, note Daly and Wilson, because "here is property that can understand its owners' purposes and weakness, that may have relatives and friends, that can plan escapes or attacks with as much foresight and ingenuity as its owners, and that can play rivalrous aspirant owners off against one another." In many respects women are successful in countering these attempts by males to control their reproduction—and men's worries about being cuckolded are as real as women's abilities to outwit them: Surveys suggest that at least 30 percent of married women have extramarital affairs. One study that examined the blood types of newborns revealed that about 10 percent of the children in North America are born to husbands who believe that the children are their own offspring, when in fact they are not. Women also not only can counter the desire of men to control their sexuality, but can occasionally use that desire to their advantage. Studies of why women have affairs, for instance, reveal that one reason women give for their infidelity is to manipulate their relationships by goading their main partners into paying more attention to them.

In the end, our growing understanding of our evolved psychology may help alleviate the ongoing sexual tension between men and women. Evolutionary psychology may suggest that it is "natural" for a man to desire sex with anonymous women, for instance—but at the same time, it suggests that it is just as natural for women to find that kind of behavior potentially threatening. If men had a greater understanding of women's evolved sexual psychology, their efforts to sexually coerce women on the street and at the workplace through lewd language and overt gestures, for instance, would be revealed not only as offensive, but idiotic. The most important prediction of evolutionary theory is that while males and females will have conflicting strategies in their sexual behavior, people will compromise those desires for the sake of being able to begin and maintain cooperative relationships for their long-term mutual benefit. The result is that an array of social and cultural traditions around the world has arisen to help cement a social contract between two people who throughout their lives will often have conflicting goals and desires.

JEALOUS RAGE

Like all cooperative ventures, the relation between husband and wife typically is based not on actual conflict per se but on the conscious or unconscious recognition by each partner that cooperating is the best strategy for gaining the most out of a relationship where people often have conflicting goals. As in the strategy of TIT FOR TAT, the potential for conflict serves as a far-distant enforcer that keeps a successful relationship going.

Unfortunately, in extreme cases such conflicts can result in lethal consequences. For unlike the TIT FOR TAT computer games, maintaining cooperative relationships in the real world is made harder by the lack of concrete information about payoffs and the limitations of the human mind to dispassionately assess costs and benefits. Being passionate plays an important role in cooperative ventures: Having a reputation for being passionate to the point of being irrational can sometimes work to one's benefit, for instance, because threats can drastically alter how each partner views potential payoffs and punishments. Threats and emotions that lead to disaster can and do happen, particularly in cases of sexual infidelity. One of the leading causes of homicide, for instance, is a male's uncertainty over paternity and the jealousy it creates.

The capacity of humans to kill others is a subject of intense debate among researchers in human behavior. Humans have long been regarded as creatures whose thin veneer of civilization coats over an ugly, violent nature that is a vestige of our ancient "animal" past. New research into the evolutionary roots of violence has shown, however, that the notion of a savage beast lurking within all of us is a myth.

Chapter Six

———

THE BEAST

WITHIN

Man, biologically considered, and whatever else he may be into the bargain, is simply the most formidable of all the beasts of prey, and, indeed, the only one that preys systematically on its own species.

—WILLIAM JAMES

The raiding party crept stealthily into enemy territory, on the lookout for trouble. The sound of a snapped twig in the distance sent a hushed wave of expectant excitement through the raiders, but it turned out to be only a bushpig rooting for food. Then suddenly the intruders heard the soft call of an infant nearby. Racing to the scene, they found a mother and her offspring trying desperately to flee. But it was too late. The raiders surrounded their victims, kicking the mother and dragging the infant through the underbrush. The attack was over in an instant: Bleeding profusely, the mother suddenly broke free, snapped up her wailing infant and fled into the brush.

The chimpanzees in Gombe had launched another successful attack.

Primatologist Jane Goodall's discovery that chimps made "war" on their neighbors helped erase the heavy, blood-red line that was thought to separate humans from the rest of the animal kingdom. Chimps had long been regarded as happy, gregarious primates who were paragons of peaceful coexistence. Humans, on the other hand, were thought to harbor a deeply ingrained "beast within" that occasionally—some people argued inevitably—reared up through the thin veneer of civilization to wreak havoc.

But, in fact, our evolutionary cousins are quite capable of committing acts of violent aggression against one another: Over a period of several months Goodall watched in horror as a group of chimpanzees she had been studying at Tanzania's Gombe National Forest systematically hunted down the members of a neighboring chimp group and killed them, one by one. Eventually they wiped out the entire community. While demonstrating that humans are not unique in their capacity for violence, studies of primates also reveal that aggressive behavior is a more complex phenomenon than the mere expression of some sort of beastly, violent drive that lurks within. Rather, aggression is often used by many primates—including humans—as a sophisticated tool that helps them cope with the challenges of their complex social world.

The idea that aggression is a vital tool for an intensely social species should perhaps not be surprising. The success of even a cooperative strategy like TIT FOR TAT depends not only on being nice but also on being willing to punish those who are not so nice. From spanking to snubbing to solitary confinement to rioting in the streets, the threat of retaliatory aggression by parents, peers, society-at-large, or an angry mob serves as an enforcer that helps make sure that the partners in cooperative ventures keep up their end of the bargain.

Ironically, it is the fact that humans are the most cooperative of all creatures that makes our aggressive encounters more dangerous than those of any other animal. In all the debate over whether humans are fundamentally naughty or nice, there is little mention of the fact that some of the greatest horrors of the world have come about through blissful, dedicated getting-along among people—from street gangs roving the back alleyways to soldiers helping each other on the front lines to a team of scientists laboring around the clock to perfect the ultimate weapon of destruction, a hydrogen bomb. The fruits of cooperation that make human society possible—planning for shared goals, marshaling of resources, sophisticated coordinating of the activities of many people, division of labor, and sharing technology—are the very same features of human society that make possible intense violence and warfare. Like those mental mechanisms that make our remarkable cooperative abilities possible, the roots of our violent behavior lie in the evolutionary legacy left by our ancient ancestors.

PRIMORDIAL URGES?

The question of whether aggression is an innate part of the human psyche has been a staple of scientific and not-so-scientific debate for centuries. Darwin suggested that humans possess a lingering stamp of uncontrollable passions inherited from our apelike ancestors and now submerged beneath modern civilization. In creatures of lesser intellect, such primordial aggressive instincts "might be necessary & no doubt were preservative," Darwin wrote in his notebooks. "Our Descent, then, is the origin of our evil passions!!"

Darwin's notion of a "beast" lurking within the human psyche was echoed in the writings of Freud, who saw aggression as "an innate, independent instinctual disposition" that was one of the most influential forces in the human intellect. The Austrian ethologist Konrad Lorenz followed Freud's lead by suggesting that while innate, aggressive instincts were necessary for survival in *animals*, in humans this natural instinct had gone terribly awry. Our "killer ape" ancestors seemed natural antecedents to a species that had just finished two world wars and was beginning to embark on an age of nuclear terror. "An unprejudiced observer from another planet," wrote Lorenz, "looking upon man as he is today, in his hand the atom bomb, the product of his intelligence, in his heart the aggression drive inherited from his anthropoid ancestors, which this same intelligence cannot control, would not prophesy long life for the species."

More recently, these bloodthirsty images of our ancestors have been eclipsed by the arguments of many researchers in the social sciences, who contend that it is not human nature that is to blame for human aggression, but civilization itself. Violence is an outgrowth of inequality, oppression, class society, and civilization's other darker aspects, argue these researchers. It is these forces, not some innate, aggressive drive, that cause the "naturally" peaceful human primate to become a raging killer. If these cultural forces could be removed, goes the logic, then violence would disappear, too. This view of our nature was recently echoed in a proclamation issued by a congress of scientific researchers who declared that "aggression is neither in our evolutionary legacy nor in our genes."

Yet, as well-meaning as such sentiments are, the entire debate over whether *nature* or *nurture* has the most influence in human affairs is built on an intellectual house of cards. Culture is not an independent entity that struggles to override human biology, but rather is a reflection of that biology, as inputs from the surrounding physical and social environment combine with the evolved mechanisms in the brain to produce the full panoply of human behavior.

The combined influence of the mind's evolved mental mechanisms interacting with the environment is demonstrated by numerous studies of monkeys and apes. Humans share more than 99 percent of their genes with chimpanzees, and studies of how these intelligent, cooperative, group-living creatures use aggression in their world can reveal much about ours. Studies of rhesus monkeys demonstrate that the social environment in which a primate is raised as an infant can have a long-term effect on its aggressive behavior. Monkeys who are reared in isolation by a mechanized, "surrogate" mother, for instance, nearly always develop into hyperaggressive social misfits who cannot get along with other monkeys.

A monkey's genetic makeup also plays a role in the level of aggression in its behavior. In a typical group of rhesus monkeys, for instance, there is usually a small percentage of males who do not get along with anyone. Like the monkeys who are raised in isolation, these monkeys disregard the group's social structure, frequently engage in aggressive acts without provocation, and sometimes are expelled from the group altogether. These monkeys are often the offspring of fathers who display similar traits, suggesting that at least part of their behavior is influenced by the genes they inherit. Though there is no specific gene for aggression, an animal's genetic makeup influences the makeup of the chemical soup of hormones and neurotransmitters in the brain and body that help mediate behavior. Hyperaggressive, antisocial monkeys, for instance, are consistently found to have a low level of a particular by-product of a brain chemical called serotonin. Conversely, high levels of serotonin are typically found in the "leaders" of a monkey group. It is thought that the higher levels of the hormone help leaders control their aggressive impulses, and studies show that it is not a monkey's fighting ability that is the key to being dominant but its ability to forge alliances with other monkeys.

Lower than average levels of serotonin also exist in men who have been discharged from the Marines for excessive violence, as well as in criminals in Finland who committed acts of wanton violence. Being born with low levels of serotonin does not inevitably lead to violent behavior. A monkey who is born with an average level of serotonin—but raised in a poor nurturing environment such as a mechanized surrogate mother—will often have a lower serotonin hormone level and high aggressiveness as an adult. Conversely, monkeys who inherit a low level of serotonin from their parents nevertheless may have their hormone level raised as the result of care and nurturing.

The complex role that aggression plays in primate life suggests that people are fundamentally neither aggressive nor nice. In the view of evolutionary psychologists, our ancient ancestors evolved to be capable of using many different strategies to meet many different goals, and that sometimes, though rarely, these strategies include threats, verbal abuse, violence, murder, and warfare. People do not merely calculate the costs and benefits of their actions and use violence whenever it suits them, however—though this can happen. More often, violence and aggression are symptoms of the fact that most human relations, from that between a man and woman to that between two bordering nations, typically contain elements of both cooperation and conflict. Our minds have evolved specializations that enable us to resolve conflicts in a way that often leads to mutual benefits. Occasionally, however, misunderstanding or miscalculation by one party or the other can lead to the relationship erupting in violence.

The finding that we have evolved mental mechanisms that help us deal with potentially violent social situations by no means dictates that humans have an innate propensity for violence. Understanding the role that aggression plays in human life and culture—and how the brain has evolved psychological mechanisms to deal with aggression—is the key to understanding the roots of violence and, ultimately, how society might bring unwanted aggression under control.

EVOLUTIONARY CONFLICT

The husband-and-wife team of Martin Daly and Margo Wilson of Mc-Master University in Toronto has been studying the link between evolution and mayhem for more than a decade. Their book *Homicide* is considered the definitive text on who kills whom—and why. "Most criminologists do not think of humans as individual actors pursuing their own interests," says Wilson. "They think of people as blank slates, unaffected by evolutionary history, and people who use violence are regarded as 'abnormal,' acting the way they do because they have been 'socialized' wrongly or live in an aberrant environment. But we are interested in looking at criminals as people who are actively pursuing strategies to further their own interest—in a sense, going out and earning their living on the other side of the law. The confrontations between young dudes in a bar or pool room, for instance, are *very* important socially—the participants are using evolved mental mechanisms that are processing all sorts of social information: For one thing, these confrontations typically occur in front of an audience of people who know them, and there are serious consequences to a person's reputation if they appear to be a chicken, or if they are successful in demanding deference from someone else. Of course, there are thousands of such interactions that are not fatal. They get resolved. The ones that wind up as homicides give us a window into human nature."

As with all evolved mental mechanisms, those that are involved in violent actions are part of every human in every culture. "When you have a model of human nature that gives you a fundamental understanding of what conflicts of interest are all about, there's no reason to expect that there will be vastly different patterns in violent behavior," says Wilson. "You have different levels of violence in different societies, of course, but the pattern of who kills whom, and the sex and age of the participants, and reasons for such conflicts, don't change. There are patterns in violence that appear across all cultures, such as violence linked with sexual infidelity and demanding deference, the fact that there is far more male against male violence than female against female; and how the risk of violence declines with age. All of these have an evolutionary history."

SEXUAL BATTLEGROUNDS

The delicate balancing act between cooperation and conflict among partners in a cooperative bond is particularly evident in relations between the sexes, where two people with often conflicting goals must balance their owns desires and actions against those of their partner—and the evolutionary stakes of being defected upon are high. The result is that the balance can sometimes tip tragically out of control. Most of us may harbor the deepest fears of being a victim of the kind of violence that typically makes front page news, such as random shootings or a robbery that ends in murder. But, in fact, in the majority of homicides, the victim's assailant is an acquaintance, friend, or spouse. Examining the police homicide records of more than 500 murders in Detroit over a period of several years, Daly and Wilson found that more than two thirds of the homicides involved people who knew each other, and stemmed from arguments over social relations, as opposed to being the result of economic crimes such as robbery.

In the majority of these socially linked homicides, the motive in these disputes was jealousy—typically *male* jealousy. An eighth of all the homicides in the United States are perpetrated by husbands murdering their wives, and in a study of murders among husbands and wives in Canada, Daly and Wilson found that 85 percent of homicides were related to sex. Numerous other studies in both technologically primitive and industrialized societies also point to male jealousy as a leading motive for homicide. The murder of a spouse is merely the tip of the iceberg of coercive violence: Jealousy is also a prime motivation in wife beating—which claims some 2 million American women as victims each year—and other forms of nonlethal violence.

Daly and Wilson do not suggest that men are blindly controlled by some sort of innate drive that inevitably leads to a violent, jealous rage with the first sign of infidelity in their spouse. The researchers argue that violent behavior arises as part of the cornucopia of different strategies, desires, and controls that men and women use to influence each other's behavior. Marriage is one of the most intimate, longest-running, and often most complex relationships that most people will enter into in their lives. Each sex brings a different set of evolutionary goals to the relationship, and there are fundamental conflicts of interest that reverberate through society and culture. One major source of conflict between the sexes is that men typically try to control women's reproductive careers—and women typically struggle to resist this male coercion—a situation that in extreme cases can erupt with tragic consequences. "There is brinkmanship in any such contest," say Daly and Wilson, "and homicides by spouses of either sex may be considered slips in this dangerous game."

Because violence is not the result of some sort of predestined emergence of a beast within, male aggression against women in response to jealousy varies from individual to individual and society to society. Evolutionary theory suggests, for instance, that aggression by men against women is often dependent on how much men and women *cooperate* with each other. In the egalitarian social structure often found in foraging and horticultural societies, males greatly depend on each other for food and protection, and so men often try to decrease the risk of jeopardizing their cooperative alliances with other men by drawing attention away from the fact that they are competing for women. In these societies, women are regarded as dangerous "spoilers" whose sexuality needs to be controlled because it threatens the group's harmony, and adultery is often blamed on a woman's sexuality—not men's behavior. Women who stray from their expected role as belonging to one man—by traveling alone, consorting with many different men, or even merely dressing provocatively—are often considered as outside the protection of a man and therefore "fair game" to other men. In less egalitarian, stratified societies, however, where some men have more wealth and power than other men, women of high status are regarded as chaste and pure—and in need of protection from low-status men, who are the ones who are regarded as dangerous spoilers.

The amount of aggression that men foist upon women in a particular society also depends on the strengths of the cooperative relationships among women. Violence against women is less pronounced in societies where women maintain strong ties to their sisters, mothers, and other women, because of the potential for retaliation, either directly or through social ostracism. Violence against women is also lower in cultures where women are more economically independent, because they have the means of escaping a violent relationship.

Since it is not in a woman's evolutionary interests to go along with a situation in which her reproduction is controlled by others, the "mating rights" over women that men often assign each other are sometimes maintained by force—resulting in violence that is often sanctioned in extremely male-dominated communities. Laws in many societies around the world regard a man's eruption into violence in response to infidelity as expected, and even excused: In some societies, other males typically do not interfere when a husband beats his wife for adultery, for instance, even if the woman who is being abused is a sister or daughter.

Adultery is typically defined in laws around the world in terms of female adventuring outside the marriage, rather than male infidelity. A woman's adultery was punishable by death in seventeenth century England; it is grounds for divorce in many modern societies, and in the not too distant past, laws in several states in the United States regarded killing an adulterous wife as no crime at all. As Daly and Wilson point out, the term "rule of thumb" arose from an eighteenth century court decree that "a husband was entitled to use a stick no thicker than his thumb to control an overly independent wife." Society's acceptance of male violence against women is fueled in part by its view of adultery as a form of trespassing against property. "There is a deep-seated psychology in men about being sexually proprietary about women—it is an absolute universal as far as I can tell," says Wilson. "Men have this bizarre feeling that women are 'turf.' Of course, women don't agree with that." Because women are viewed as property, violent husbands are seen as engaging in a kind of protective behavior similar to self-defense. Women must deal with conflicting evolutionary strategies of their own: A women may view another sexually active woman as a rival who threatens her relationships, for instance, and therefore will sanction the punishment of adulteresses—even though she herself may desire to have a variety of mates.

SUFFER THE CHILDREN

The dangers posed by a male's evolutionary concerns about a woman's sexuality ultimately extend to children as well. Since in every human society, males contribute resources to the raising of children, evolutionary theory suggests that males would be very keen to make sure that any infant they help raise is their own. The male's concern over paternity is evident in how people respond to newborns: Transcribing the remarks of a sample of mothers in North America immediately after childbirth, for instance, Daly and Wilson found that within fifteen minutes most mothers had commented on how much the child looked like the father. Anthropologists have found that remarking on the similarity of a child to the father is a common cultural practice in many societies around the world.

Not only will fathers be especially concerned about confirming the paternity of a child, but fathers may also behave less favorably toward children that are not theirs. Among gorillas and some species of monkeys, for instance, nearly 40 percent of the infants present or born shortly afterward were killed by a newly dominant male. Human children, too, appear to run a higher risk of injury from a stepparent than from their natural parents. In one study of a foraging society in Paraguay, for instance, nearly 50 percent of the stepchildren died by the age of fifteen, whereas less than 20 percent of children raised by both natural parents died. Studies of child abuse in modern North America reveal that a stepchild is 100 times more likely to be fatally abused by a stepparent. In one year in the United States, for instance, nearly *half* of the 276 victims of fatal child abuse lived in families with one stepparent—a pattern of violence that is unrelated to the parents' socioeconomic status, age, or family size. Small wonder that in folk tales around the world, stepparents are commonly portrayed as villains who pose a danger for children.

As with all studies of evolutionary psychology, identifying an evolved psychological mechanism that influences behavior does not necessarily mean that all stepparents, for instance, are destined to be violent abusers of their children. "Homicide is merely a window on the emotion, motivation, information processing, and so on, in how parents care for their own children versus caring for someone else's child," says Wilson. "We argue that there *must* be a parental psychology. Mainstream psychologists typically don't say that: They say that parenting is only a role—if you say your lines well, then you are just as good a parent. But to us, that just doesn't capture the intensity of parents' commitment to their children. So we say that selection has shaped psychological mechanisms in the brain that manifest themselves as a *tremendous* commitment to a child—and not just any child. This belief that there is a parental psychology prompted us to look for lapses of it. And the only reason we look for lapses is that you can get data for the lapses—the homicides and abuse. No one keeps records on the vast majority of stepparents who are good to their children."

If evolutionary psychology predicts anything "innate" about people, it is that their minds are exquisitely crafted by evolution to form cooperative relationships built on mutual trust and kindness. Still, the startling statistics on child abuse suggest that the stepparent/stepchild relationship exists on a different grounding from that of natural parents, and so is more vulnerable to breaking down under pressure. It is not that parents consciously try to protect their own genes by being less abusive to their natural children—after all, there is no difference in homicide rates for adopted and biological children. The evolutionary psychology involved is more likely to be similar to that of incest avoidance: The experience of living with a child from the time of its birth triggers evolved psychological mechanisms that cause a particularly strong bonding with the child—something that is perhaps not as strongly developed in a stepparent who begins to live with a child when it is older.

SAME SEX COMPETITION

The model of human nature found in mainstream social science—which regards male and female minds as identical and infinitely plastic—does little to explain the fact that, all over the world and throughout history, the vast majority of violent acts have been committed by men. In North America, for instance, males commit more than 80 percent of homicides, most of them against other males, and in other cultures the percentage of males committing murder goes as high as 99 percent.

Around the world, the most common culprit in committing a murder is a young man in his teens or twenties, with the rate of murder quickly tapering off as men get older. "When we first looked at these data on the computer I couldn't believe the similarities," says Wilson. "If you take these curves from the various countries and lay them on top of each other, they look like just one line." Even though the actual rates of murder are different among different societies, the age and sex of the murderers are the same the world over. This sharp rise in murder rate among postpuberty males coincides with the peak of their physical prowess as well. It is in a man's late teens and early twenties that he has the greatest aerobic power and physical strength, and the quickest reflexes, suggesting that the development of the male physique coincides with the period in his life when he is most at risk for getting in social conflicts that potentially lead to violence.

Evolutionary psychology helps explain why men at this time of their life commit the most murders. "We are trying to find out what mediating factors might make it likely that a male would engage in the kind of behavior that would make him part of these statistics," says Wilson. "And the first thing you suspect is that the fruits of success in such a confrontation—status, reputation, resources—might be more valuable for men than women at this stage in their life." Most homicides are not over material resources such as money, cars, drugs, etc. "Rather, it is the importance of reputation for fierceness, for not backing down, for demanding—and getting—deference, that is what *a lot* of these disputes are about," says Wilson. "These homicides most often occur in front of friends of the victim and the murderer—they have an audience—and so there are real social payoffs to winning such a confrontation."

Daly and Wilson's ongoing studies of who commits murder—and why they do—suggest that homicides are often the outgrowth of disputes over things that evolution has crafted the human mind to be quite concerned about, such as status, reputation, and, ultimately, access to mates. An evolutionary psychology perspective suggests that there should be more competition between members of the same sex than between members of the opposite sex, for instance, because members of the same sex often compete for the same resources—and one of the most important of these resources is members of the opposite sex. In those societies that are stratified into haves and have-nots, a high-status male can monopolize the sexual attention of more than one female—meaning that low-status males face the possibility of having no mates at all.

The competition for access to females—and the prospect of an "evolutionary death" if a male is shut out by other males—results in low-status males having less fear of biological death. Thus low-status men would be expected to take more risks in situations where there was a potentially high payoff in terms of sex, status, or resources. Such high-stakes games are fertile grounds for violence: Homicide data reveal that murderers and their victims typically belong to groups with low economic and social status. "In gangs, for instance, the risk of dying through violence is high," says Wilson. "But they have no other ways of controlling their conflicts: They don't have access to the courts, they don't have resources, they can't give benefits to anybody. If someone cheats on a contract, they have to exact their revenge by imposing the cost themselves. They monitor cheating and defections very, very closely, and they use the occasional death as a very effective deterrent."

An evolutionary psychology perspective suggests that in a tragic, but very real sense, there is nothing "abnormal" at all about the level of violence among gangs—given the level of poverty, education, and opportunity among the people involved. "If you have something, then to lose your life over it isn't worth it," says Wilson. "But if you have nothing, then death, lifelong celibacy, and zero social status are equivalent. From an evolutionary perspective, you *ought* to have a mind that devalues your own life if your expectation of reproductive success is zero. And the more you devalue the future, the more risks you are willing to take." The causes of the social conditions of many murderers and their victims are no doubt intricately tied to the class and race problems that plague modern society today, but the result is as old as the human species: Throughout human evolution, men with low status and few resources have run the risk of attracting few mates and hence few offspring, and so were more likely to take dangerous risks to boost their chances—risks that include situations that can lead to violence.

The importance of status and reputation among men helps explain the finding that many homicides stem from seemingly "trivial" arguments. In one study of 560 murders in Philadelphia, for instance, Daly and Wilson found that nearly 40 percent of the homicides were attributed to disputes over curses, insults, and one-upmanship. Status and reputation are important in every socioeconomic class. But men of low status have more to gain and more to lose in public displays and defenses of their reputation. While many homicides may appear to be caused by trivial disputes, Daly and Wilson's study of similar cases in Detroit has convinced them that something important *is* at stake: Violent male-male disputes are really concerned with "face," dominance, and presentation of self.

The propensity of some encounters to lead to violence is made more likely by the fact that, while the risks of becoming a victim in a violent standoff may be high, the legal consequences of "winning" such a dispute are often not severe. Following up on the homicide cases they studied, Daly and Wilson found that out of 121 homicides where a suspect had been identified and apprehended, nearly half of the cases never went to trial or resulted in acquittals. Only 14 of the cases that resulted in convictions were actually for first- or second-degree murder; in the other 50 cases the convictions were for manslaughter or a reduced charge, where the convicts could be paroled after 18 months. Daly and Wilson suggest that this inability of society-at-large to effectively punish crimes can lead to a pattern of retribution by the victim's family or alliances, which in turn leads to further violence: Many urban homicides, for instance, are carried out by gang members who are exacting revenge for previous killings of people who were in their gang.

CALLING A BLUFF

The ultimate deal-making machine, the human psyche has evolved mechanisms to deal with the subtle ins and outs of negotiating threats of violence. In a threat such as "If you eat my food I will kill you," the terms of the arrangement are true in reverse as well: If the threatened person does not eat the food, he or she will not be killed. Thus if you are threatened, you must be wary of two possible outcomes: One is that the threat is a bluff—the person won't kill you even if you do eat the food. The other potential outcome is a double cross, where the person kills you anyway even if you comply with the threat and don't eat his food.

The differences in physical strength between males and females have resulted in the male and female mind evolving different abilities in detecting bluffs and double crosses. Since men are typically more equally matched in a standoff, it is more dangerous for the threatening man to actually carry out his threat against another man. Therefore, one would expect males to bluff other males more than they bluff females, with whom they have an overwhelming advantage. And since it would be costly for a man to give into a threat if the threatener is in fact only bluffing, the male psyche evolved to be keenly on the lookout for the possibility that someone who makes a threat is bluffing. At the same time, one would expect that women would be less concerned about detecting bluffs from a man in a threatening situation, because most of the time, a man can easily carry out his threat against a physically weaker woman, and so women are subject to fewer bluffs. When it comes to detecting a double cross, however, a woman should be just as wary as a man, because both males and females have much to lose if a double cross is carried out.

Probing the minds of males and females with a series of logic puzzles, John Tooby and Leda Cosmides found evidence that men and women do appear to have different, specialized mental mechanisms that have evolved to help them deal with threatening situations. For this experiment, the researchers used a test similar to the test Cosmides used to reveal our evolved mental mechanisms for detecting cheaters. In this case, Tooby and Cosmides returned to the fictitious Kaluame people, and created a scenario involving a man named Big Kiku threatening, "If you eat any of my cassava root, I will kill you." They presented four cards reading *eats food*, *kills you*, *doesn't kill you*, and *don't eat food*. Tooby and Cosmides found that when the scenario was written to portray Big Kiku as having a reputation for being cruel, and people were instructed to look for the possibility of a double cross, a majority of both men and women performed equally well—turning over the cards *don't eat food* and *kills you*. As in Cosmides' test of cheater detectors, these people performed better than those subjects who were given the abstract version of the test asking simply to match letters with numbers—further evidence that the minds of both men and women have been designed by evolution to reason better when they are in social situations, such as when they must be on the lookout for double crosses. True to the predictions of evolutionary psychology, Tooby and Cosmides also found that women performed worse than men in detecting a bluff. In this case, the scenario was recast so Big Kiku was portrayed as a coward, and people were instructed to look for evidence that he was bluffing—which would require turning over the *eats food* and *doesn't kill you* cards. In this case, men performed better than women.

KEEPING A PRIMATE PEACE

One of the key roles that aggression plays in a society is, ironically, to keep the peace. Aggression helps maintain the complex pecking order that typically characterizes primate groups. In rhesus monkeys, for instance, both males and females have their own separate "dominance hierarchy," with most individuals being both directly above and below someone else on the ladder. While being a dominant doesn't necessarily guarantee exclusive rights to food or mates, such hierarchies typically dictate who among a group will have first access to essential resources. With such hierarchies, the question of who goes first is settled beforehand, which helps to keep day-to-day aggression among individuals to a minimum. Studies show that encounters in which a monkey peacefully defers to a more dominant monkey occur three times more often than incidents involving overt aggression between the two. Likewise, dominant males rarely use the full power of their strength and their large, sharp canine teeth to deter challenges by others, depending instead on threatening displays to make their point.

The importance of the group's hierarchical social structure in keeping the peace is evident when a newcomer tries to join the group. In most primate species either males or females migrate out of their local group to join another. In rhesus monkeys, it is young males who must make the switch as they come of age—typically at the insistence of the dominant males. When the young monkey tries to join a new group, his presence results in a jostling of the social hierarchy as he tries to work his way into the social ladder, and this often leads to fighting. It is an extremely difficult time for young male monkeys: As many as half of them die during the process of transferring to a new group—either through violence from members of the group or because they were prevented from joining, and are killed by predators or die of starvation. Once the monkey is established in the group, however, the level of aggression toward him drops dramatically, as each individual generally defers to those above it in the societal ladder.

Aggression plays a key role in the mutual give-and-take that is essential for maintaining cooperative alliances. In one instance, a chimp who was under attack extended his hand toward another chimp in an apparent plea for help. The chimp under attack had come to the aid of the other chimp during a previous fight, but his request for help was ignored. In a classic example of playing TIT FOR TAT, the chimp immediately turned and attacked his betrayer for his "defection" from their previously cooperative relationship. Other studies suggest that aggression within a group is often focused on those individuals who act selfishly. In one experiment, a bundle of food was placed among a group of chimpanzees. As the chimps jockeyed with each other for access to the food, the most aggression was directed not against those chimps who would not share their bounty but against those who had not shared on a *previous occasion* when food was delivered and were now asking others to share. By policing cooperative relationships, aggressive behavior helps foster cooperation.

THE ROOTS OF WAR

Ironically, it is our species' supreme powers of cooperation that make war possible. Chimpanzee group-against-group conflict is similar in principle to warfare among some small-scale human societies: Both are carried out by coalitions of males, both are the result of imbalances of power among neighboring groups, and both have their ultimate roots in males' desires to gain access to females. In the case of the group-group conflict among the chimpanzees at Gombe, the "aggressor" chimps took over their victims' territory and incorporated several of the females who were attached to the other group into their own community. A comparison of the warfare patterns in forty-two foraging societies worldwide reveals that when people live near vital resources such as fertile land or watering holes, the group-against-group aggression is typically over control of these physical resources. Yet, ultimately, these resources are also tied to access to females: Those individuals in the society who accumulated the most wealth typically had the most wives, and so gaining resources is an important factor in access to women.

When there are no defensible resources to fight over, group-against-group conflicts often directly concern disputes about access to women. For instance, among the Yanomamo, a group of some 15,000 people who live in about 200 villages in the Amazon rain forest, a typical fight starts when a man from one village abducts a woman from another village for a wife. According to anthropologist Napoleon A. Chagnon of the University of California at Santa Barbara, who has conducted a lifelong study of the Yanomamo, the woman's relatives will retaliate by launching a revenge attack against her abductors. Each attack invites further retribution by the victim's relatives, which is made more likely by the fact that in a typical Yanomamo village nearly everyone is at least distantly related to each other in multiple ways. The result is that "blood revenge" may escalate in a cycle of conflict that can last many years. Because a village's reputation for swift retaliation serves as a deterrent for attacks from other villages, the cycle is perpetuated.

Despite the tragic, long-term damage that results from such conflicts, some Yanomamo men do reap rewards for taking part in the aggression. Successful Yanomamo warriors are awarded a high-status position in the community, which typically translates into having more wives and more children. Yanomamo men who have killed someone else in battles have an average 2.5 more wives and three times more children than those who have not killed. Yanomamo warriors do not consciously strive to kill people so that they can have more wives, nor do they possess some gene that makes certain people more warlike, which gets passed on to their offspring. Rather, a man's reproductive success is tied with his status in the community, and in Yanomamo society status is closely linked with willingness to defend, often violently, one's kin or members of the community. In this respect the Yanomamo are little different from people in industrial societies, where men in urban gangs play a violent, high-stakes game for status, and political candidates promote their military record—or their opponent's lack of one—as evidence of their worthiness for office.

Group-against-group warfare is not solely a male trait in primates: In those monkey societies where females, not males, form strong cooperative alliances—as is the case in most nonhuman primates—it is the females who typically band together to fight aggressively over resources. But in humans and our closest evolutionary cousins, the chimpanzees, it is males who stay at home as they mature, and females who migrate to another group. The result is that in both chimpanzees and humans, males form intense alliances—"old boy networks"—which they use to compete against other groups of males. While in humans, females form more complex alliances than in any other primate, it is nevertheless still a distinctly male trait to form groups for the purpose of attacking another group of the same sex. Women may serve in the military, but nowhere in the world have women formed armies to fight other groups of women for access to men.

Primate studies suggest that human warfare is not merely a product of some inevitable, innate, "beastly" urge for violence, nor is it merely some sort of pathological reaction to the pressures of modern civilization. Ironically, the biggest factor in triggering group-against-group conflict in both chimpanzee and human foraging societies is *cooperation:* In chimps and technologically primitive human societies, a group typically attacks another group only if their side vastly outnumbers the other, so there is little risk of physical harm to the attackers. The reason that human warfare is more lethal than the fights among chimpanzee groups is not that humans are more violent than chimps, but that humans are more cooperative and so are capable of unleashing more power in their expressions of violence.

It is this ability to form close-knit coalitions and alliances among a group that makes possible violent attacks on other groups. In many ways, cooperation within a group—and the competition between groups that this cooperation makes possible—are both facets of the same evolutionary strategies: Just as the human mind has evolved mechanisms for maintaining group cohesion, it has also evolved a keen sensitivity toward rooting out potential "defectors" and a distrust of passing strangers—for whom the potential for long-term cooperative relationships is low. It is no mere coincidence, therefore, that those species that are most capable of group cooperation are also those who are most likely to engage in group-against-group aggressive conflict.

FROM INDIVIDUALS TO GROUPS

Whether they are small bands connected by blood relations or large-scale communities bound together by ethnic ties or nationalism, societies are continually engaged in a huge, multiplayer cooperation dilemma. Like the cooperative relationships between individuals, interactions between groups are prone to the same kinds of problems in maintaining a relationship, such as working with incomplete information and having decision making muddied by conflicting long- and short-term goals. Warfare, like many acts of violence, typically represents a breakdown of a cooperative exchange because of miscalculation by one party or the other—few nations willingly go to war expecting to lose. As demonstrated by Poland succumbing to Hitler's army and Kuwait's rapid occupation by Iraq, many nations choose not to fight when it is clear who the winner will be. But when it is unclear who will prevail the prospect of having no future leads to a cooperative breakdown that often turns into war.

The fact that leaders may have personal goals that may conflict with those of their country adds a dangerous spark to international relations. In large, complex societies, the potential for the breakdown of cooperative relations between nations becomes greater, because typically the decision to go to war is made by a small group or single leaders in the societies. Despite the complexity of their society—and the fact that thousands of lives may be at stake—leaders may still be strongly influenced by concerns such as "saving face," personal ambition, and maintaining their reputation. The situation is complicated by the fact that in state societies, the people with the decision-making power to enter a war are typically not the people who directly pay the cost of waging it. As might be expected, this drastically changes any cost-benefit considerations a leader might undertake before deciding to fight.

One encouraging finding in the study of warfare is that the more people participate in their government, the less likely they will be to engage in war. For nearly two centuries, no democratic nation has waged war on another democratic nation—though, of course, they have fought against nondemocracies. This relationship between the level of popular participation in government and the state's willingness to wage war appears to hold for nonindustrial states as well. The findings suggest that the more people involved in a political decision, as in a democracy, the less likely there will be a breakdown of the cooperative bargaining between nations, and war.

Crucial to people's participation in the political process, as it is in nearly every other part of our everyday life, is the uniquely human ability to use language. One of the suite of evolved mental mechanisms that help create the dense web of social life that characterizes our species, language is not the only thing necessary for social relations. After all, our evolutionary cousins form intense social bonds perfectly well without it. Yet language allows the rest of our evolved mental machinery to express itself to the fullest and in a way that is uniquely human: With our ability to speak our minds, the simple songs of primate behavior become a raging symphony.

Chapter Seven

———

GIVING THE

MIND

A VOICE

Language is a form of human reason and has its reasons which are unknown to man.

—CLAUDE LÉVI-STRAUSS

B y her first birthday my daughter had learned that things could come out of her mouth as well as go into it. First came the names by which she could summon her parents; then the sound those parents would no doubt hear thousands of times again over the years: "*NO!*" Month by month into her second year she has added new words like pearls on an infinite strand; soon she will be rearranging those pearls again and again to tell the world not only what she has on her mind, but, more important, that her mind is a distinctly human one.

Language is the calling card of our species. As far as history shows, all human societies have had language, and all humans have the potential to communicate by speaking, writing, and signing. Children are equally capable of learning any language, from Arabic to Zapotec, and are fluent by age three with little or no instruction from others. In some cases children have been shown to invent their own full-fledged languages— called *creoles*—from different bits and pieces of various languages they hear from adults around them.

No mere cultural invention, language is a natural property of the human psyche. Though there are some 5,000 different languages in the world, each with its own vocabulary, inflections, and quirks of grammar, the basic human capacity to speak and understand language has nothing to do with the cultural trappings of any particular society. Language was not invented at one place and passed on to others, as was often the case with agriculture and the wheel. Nor does language in any way reflect a society's political system or level of technology: The language of the !Kung San hunter-gatherers of Africa's Kalahari Desert is as complex and sophisticated as the Queen's English. Likewise, the fourteenth century English of Chaucer's *Canterbury Tales* is every bit as complex as the twentieth century tongue that gave rise to Hemingway's *A Farewell to Arms*. As the noted American linguist Edward Sapir once put it, "Linguistically, Homer walks with the swineherd of Macedonia, Confucius with the headhunters of Assam."

IN SEARCH OF LANGUAGE'S PAST

The question of how, when, and why language arose in the human race has been debated for centuries. "The study of the evolution of human language has always had a bit of a bad reputation," says Paul Bloom, a linguist at the University of Arizona. "In 1866, the linguistic society in Paris banned the whole discussion of the topic. And many academics today tend to view the issue as pretty much a fringe, eccentric kind of research—people who spend too much of their time on it are regarded as crackpots." Bloom and his colleague, linguist Steven Pinker of the Massachusetts Institute of Technology, are two such "crackpots." Both are respected scholars—Bloom and Pinker are well known for their research on how children learn language—yet both young researchers have been lured by the power of evolutionary psychology to try to understand how we acquired our gift of tongue. "In a sense, our goal is incredibly boring," says Bloom. "All we are doing is arguing that the evolution of human language is no different than the evolution of color vision in primates, or echolocation in bats—*not* like cultural inventions such as agriculture—and so it can be explained in the same way. Once people start doing this, studying the origins of language might turn into a more respectable enterprise."

Many past attempts to explain the origins of language have been "just so" stories—speculations that were impossible to confirm one way or the other. Researchers have proposed that language arose from early humans' imitating animal sounds, for instance, or from mimicking the sounds of nature, or making grunts of exertion. These hypotheses are now derided as the infamous "bow-wow," "ding-dong," and "heave-ho" theories. "These ideas arose from the assumption that language is a cultural invention and doesn't have anything to do with brain evolution," says Bloom. "The idea was that people were walking around, not using language, then all of a sudden somebody heard a dog bark and said 'Hey, we could do that, too!' It's one of those bizarre scientific theories that portray us, as animals, lacking something commonly found in most other animals. Of course, the people who made up these stories assumed that we descended from Heaven."

But even today, when the notion that humans are a product of evolution is a fundamental tenet of all modern science, the human ability to use language is often regarded as somehow excepted from the process of biological evolution. One linguist recently asserted, for instance, that the only way to argue for a biological underpinning of our language abilities is to accept that it "was endowed to us directly by the Creator, or else our species has undergone a mutation of unprecedented magnitude, a cognitive equivalent of the Big Bang."

But Bloom and Pinker believe that just like any other complex, extremely useful part of the human body—such as the eye—language in fact did evolve. "To make our argument, we offer a classic bit of evidence that has almost been entirely ignored by scholars looking for the origins of language," says Bloom. "That's the psychology, physiology, and genetics of language itself."

A SOCIAL TOOL

Birds, bees, and monkeys in the trees communicate with each other through sound and gesture, of course. But as with so many animal behaviors, our species has taken this simple communicative skill common to many other animals and given it manifold variation that makes our language abilities unique in the animal kingdom. In keeping with the old "man the hunter" image of our ancient ancestors, language has long been thought to have arisen to help people coordinate a hunt or talk about what kinds of prey are present. Yet it is more likely that language is most useful in the activity that most distinguishes our species—being social. "The first answer to why we have complex language, and no other animal does," says Bloom, "is to point out that this attitude is anthropomorphic: Evolution has given many species innate communication systems, and human language is merely one of them—albeit a *very powerful* one. So the real question is, Why is it so powerful a system for us, and not for other animals? The answer to that question rests in the ways in which we are special: We are extraordinarily social."

Our complex linguistic abilities aptly mirror the complexities of the Byzantine social relationships that characterize our species. Language is the ultimate tool for cutting deals: It can be used to make one's goals and intentions explicit to others, offer promises or threats, or lay out the terms of a *quid pro quo*. It makes possible the rallying cry of war parties and the measured words of peace treaties. Language can also be used to alter another person's intentions through begging, pleading, wooing, coercing, and threatening. "We use language to negotiate, to persuade, to deal with one another," says Bloom. "We need a communication system to maintain long-term partnerships: Much of our life is like a 'Prisoners' Dilemma,' and if you and I are going to surmount the dilemma, we need to interact over and over—and language helps. If you and I aren't going to hang around together too much, there's not much pressure to develop a communication system of any sort."

Language is not necessarily a prerequisite for forming complex social relationships—just as important are the specializations of the human psyche such as memory for faces, reasoning, "cheater detectors," ability to project scenarios into the future, and capacity to form a theory of mind. Rather, language is grease on the wheels of social exchange, helping to make these enormously complex interactions run smoothly. Monitoring the conversations among men and women at a cafeteria, researchers found that more than half the time the subject of talk was purely social and not involved with the passing along of information. "We have social skills that are independent of language," says Bloom. "But given that we have these social skills, language suddenly becomes extraordinarily useful. And there is coevolution: Once you have social skills, and then you get language, suddenly your social environment changes radically—which increases evolutionary pressure for more social skills."

The power of our ability to use language lies not simply in our ancestors' evolving the ability to speak—after all, even parrots can produce sounds that closely resemble human speech. A more important clue to the power of language is what humans talk about: Humans do not use language simply to interact with others in the here and now. Language allows humans to move out of the present and away from their immediate surroundings. With language, humans mark places in time—past, future, and imaginary—and even mix them together, as in "You picked up the check for lunch last week. Why don't I pay for the meal today and when we get together next week we'll split the bill?"

The ability to use language to mark subtle differences in time must have been vital for our ancient ancestors. There is a big difference between saying, "If you give me some of your fruit I will share meat that I will get," and "Give me some fruit because of the shared meat that I got," and "If you don't give me some fruit I will take back the meat that I got." "Some linguists think that the complexity of language is only necessary in the hustle-and-bustle industrial world, but not for hunter-gatherers," says Bloom. "But when you look at how language is actually used among modern hunter-gatherers, you realize that language is far more important to them than to us in the modern world. I mean, I could sit alone in my room for two days and not use language. But for a hunter-gatherer, there is a tremendous social pressure for language. Their lifestyle is all making agreements, wheeling and dealing, cooperating, discussion, etc. The lives of our ancestors must have been one long encounter group."

For both men and women, language is part of the social "grooming" that occurs among many species of primates. In nonhuman primates grooming ostensibly involves painstakingly combing through another individual's hair for bugs and such, but it plays a bigger role in cementing the social bonding between two individuals. With our ability for language, however, people can use conversation to "groom" several people at once, which is part of the reason that humans are capable of forming larger, more cohesive groups than any other primate.

Both men and women use language as social tools, though slightly differently: For men, using language is part of their striving for status and building a reputation within the social hierarchy; women, on the other hand, typically use language to cultivate connections with other people, for which talking about their social interactions and those of others—gossip—forms an integral part. Indeed, women's cooperative ties to other women, in part made possible by language, are one of the key features of the human species that distinguish us from other primates. Like our close evolutionary cousins the chimpanzee and gorilla, human females in horticultural societies—and no doubt among our ancient ancestors, too—typically migrate out of the group they were born in. But unlike chimpanzees, where females lead fairly solitary lives, women form dense social connections of their own with other women. Men, too, of course form social connections with each other, but these are typically hierarchical alliances that are roughly similar to those of the males in chimps. Only female chimpanzees in captivity form similar social alliances, whose purpose might be best described as enhancing their own well-being. They make friends not to cooperate toward some goal, but for their mutual pleasure. It may be more than a coincidence that women, for whom language plays such a vital role in forming these cooperative alliances, typically develop their language skills earlier than men as children and outscore men in verbal abilities as adults.

Language not only allows us to step out of time but also out of space. With language, it is no longer necessary to directly observe people's behavior to get an idea of how trustworthy or reliable they are as potential cooperators, for instance. Language makes possible the concept of *reputation*, as word of mouth, gossip, and testimonials from others shape our opinions of whether someone would be a worthy partner or formidable enemy, without our actually having the experience first hand. Indeed, gossip and reputation may be some of the crucial functions for which language arose.

The ability to convey important information about the world without actually having to experience it is a key part of human culture. With language, a tutor can explain how to do something without actually having to do it—a valuable tool in teaching, for instance, how to avoid being run down by a mastodon or what to do if a nearby volcano suddenly erupts. While in some other species, one animal will train another, ours is the only species that explicitly practices pedagogy, that is, observing another's behavior, judging his or her competence, and stepping in to correct it. While language is not necessary for teaching, it is obviously a significant addition to it.

Most important, because knowledge can be translated into language, valuable learned information doesn't pass away with its possessor but lives on as part of the culture through lore, songs, cultural practices, and writing. Once part of culture, a particular bit of knowledge can be debated, considered, and improved upon by others. By making learned knowledge part of each person's surrounding environment and not just the inner world of the mind, language expands the capacity of human memory to include the memories stored in the culture at large. Language plays an important role in creating this collective community memory, making possible an oral or written history that connects people to their past. With language, valuable learned behaviors no longer have to be stored solely among the neural wiring of the individual brains that possess them. Language becomes a way to transfer information across generations, allowing a person today to pick up a book and communicate, for instance, with a Greek philosopher who lived 2,000 years ago.

THE ANATOMY OF SPEECH

So essential is language to the human species that each year some of us pay a dear price—death—for our linguistic abilities. Thousands of people die annually from choking on food, a direct result of anatomical changes in the throat, unique to humans, that allow us to produce the broad range of sounds used in speech. As Charles Darwin noted in his *Origin of Species*, it is a "strange fact that every particle of food and drink which we swallow has to pass over the orifice of the trachea, with some risk of falling into the lungs."

The evolutionary development that led to the need for the Heimlich maneuver is that, unlike all other mammals, humans have their larynx placed well down their throat, near the Adam's apple. The larynx is a valvelike organ that sits on top of the windpipe; in other mammals its primary function is to block food and liquids from falling into the lungs. In most mammals the larynx resides higher up the neck, near the base of the skull, which enables animals to drink and breathe at the same time—something humans can't do. Infants are also born with their larynx placed high up in the throat, but as a child develops, the larynx migrates downward toward its adult position. By three months the larynx has moved most of the way down, but the process is not complete until adolescence—which is when boys have their voice "drop" to its adult pitch.

The larynx's function as a blocker of food and drink still takes place when humans cough, but for the most part, the larynx has been co-opted by evolution for another purpose: giving humans the ability to produce the broad variety of sounds that characterize speech. If the typical mammalian vocal tract is like a bugle, the human throat is like a trumpet: When humans speak, air from the lungs is forced upward through the larynx's vocal cords, which rapidly contract and loosen to create puffs of air that are shaped into the sounds of speech. Consonants are created by manipulating the mouth and tongue, but without the extension of the vocal passageway made possible by the lowering of the larynx, the complex tones of the vowels *i, u,* and *a,* which are the most common vowels in many of the world's languages, could not be produced. Without this customized human vocal tract, human speech would be a high-pitched, nasalized whine that lacks these vowels, and understanding between people would be lowered as much as 30 percent.

Though the most important parts of human language abilities—the brain and larynx—don't fossilize, faint clues to when these physical changes for language arose nevertheless lurk in the bones of our ancient ancestors. Examining the bumps and dimples left behind on the skulls of the ancient human ancestor *H. habilis,* for instance, researchers have found an indentation that appears to have been left by a brain structure known as Broca's area. First detailed in the 1860s by the anatomist Paul Broca, this tiny chunk of brain tissue orchestrates the muscle coordination to produce speech in modern humans, and the hints of its presence in the brains of *H. habilis* suggest that the hominid may have had at least some rudimentary language abilities.

The overall shape of the skull can also give clues to our linguistic past. The base of the skull must flex downward to accommodate the lowered larynx of the fully developed human vocal tract, and only human skulls display such downward flexing—the base of ape skulls, for instance, is flat. This downward flexing in a skull gives clues to the time when full-fledged speech first appeared in human ancestors: In even the oldest known fossils of *Homo sapiens*, which date back nearly 100,000 years, the base of its skull is fully flexed, just like that of modern humans, suggesting that these ancestors could produce a full range of speech sounds. Going backward through time, the skull is flexed somewhat less in Neanderthals, the humanlike creatures who populated Europe between 100,000 and 35,000 years ago. The skull base is less flexed in the ancient human ancestor *Homo erectus*, and even less in the more ancient *Homo habilis*. The skulls of the *australopithecines* such as "Lucy," on the other hand, show no such flexing, suggesting a more apelike vocal tract.

The sophistication of the human vocal tract is yet another example that underscores the fact that researchers must look at our ancient ancestors not as isolated "Robinson Crusoes" battling alone against a harsh environment, but as an integral part of a larger network of other humans who were deeply dependent on their social relations with each other for survival. Certainly, a creature does not necessarily need a lowered larynx to produce language—people speak through sign language, for instance. But if someone is trying to convey information over a long distance, in the dark, in the midst of a firefight, or at a crowded cocktail party, speed and accuracy are essential, and having a lowered larynx that can produce a cornucopia of sounds is very helpful. In other words, while having a lowered larynx is not necessary for language, once the ability to produce language got started, it is likely to have fueled the evolution of the uniquely shaped human vocal tract to make our ancestors' language ability even more powerful.

THE ROOTS OF LANGUAGE

Of course, our ability to speak a language depends more on the human mind than the larynx. There is growing evidence that language is produced by an evolved mental "organ" in the brain that has been specially designed for the task. "The typical 'origin of language' scenario suggests language is somehow tied to other features of our large brains, such as motor control or having a hierarchical nerve structure," says Bloom. "But these explanations at some point have to be wrong, because there are people who have brains that are perfectly good at these things who can't learn language. The converse is also true: You can have language without having a big brain." For instance, so-called nanocephalic dwarfs, who are victims of developmental problems that result in their brains being abnormally small, suffer from diminished intelligence and other mental abilities but have sophisticated linguistic abilities. Another disease known as William's syndrome results in people being severely mentally disabled, with an average IQ of about 50. Yet their speech is fluent and syntactically perfect.

Having an intelligent, large brain that can do many other mental tasks well does not necessarily result in having language abilities. Several people in one particular family, for instance, have been found to suffer from a rare genetic defect that produces specific deficits in their language abilities, including the inability to understand future and past tense and to form plurals. People with this linguistic defect have otherwise normal intellectual capabilities: One boy who had the gene defect was one of the leaders in his class in mathematics and computer programming. "They have no problem learning mathematics, how to play chess, how to use a computer, how to establish relationships, how to do square dancing, whatever," says Bloom. "They just can't learn language: Despite intensive training, they can't understand that *dogs* stands for more than one dog, and *dog* is for a single dog."

The idea that language ability is an innate property of our brains has long been championed by MIT linguist Noam Chomsky. Perhaps the most influential linguist of this century, Chomsky has altered the particulars of his theories about language that he first outlined in the late 1950s. But the essence of his argument remains the same: All human languages share a similarity in their structure—what Chomsky and his followers call a Universal Grammar—and this similarity among the world's languages arises from the fact that language is produced by a brain "organ" common to all humans—an organ that was specially designed for producing language. "A human language is a system of remarkable complexity," Chomsky once observed. "To come to know a human language would be an extraordinary intellectual achievement of a creature not specifically designed to accomplish this task."

While the world's languages might superficially appear to be very different, their fundamental structure is constrained by the evolved biology of the human brain. "When you combine Chomsky and Darwin on language, it gives a real shock to our intuitive notions," says Bloom. "For instance, it is hard to tell my grandmother that languages vary only superficially—she'll think that you're crazy: What could be more different than English and Urdu? You can't understand a word these guys are saying, and they can't understand what you are saying. But in the big picture, the differences between English and Urdu are all minor variations. From an evolutionary point of view, we all speak the same language."

As an analogy for how the many different languages found around the world can arise from the same fundamental structure in the biology of the human brain, consider a pedestrian taking a leisurely walk across town to visit a museum. On any particular day he might choose one of any number of possible routes, depending on the weather, traffic lights, whether he wants to run other errands along the way, how much time is available, etc. As he walks along, each turn he takes on the way limits some future choices he might make—bookstores that won't be browsed or busy streets he must cross, for instance—while it opens up new avenues to follow, such as a walk through the park or the shortcut up the alley. The possible routes may seem endless, but there are some very real limitations to what paths the pedestrian can take: He cannot jump over apartment buildings, swim across the river, tunnel through walls, or run into oncoming cars. In this analogy, the layout of the city is like the fundamental structure of language that is part of every person's brain: While there are many possible routes to take, all are ultimately constrained by existing pathways that are already in place before the journey is begun.

A child undergoes a similar "journey" as he or she learns a language: Though there are many possible paths to follow in terms of grammar and vocabulary, the basic structure of any human language already exists as an innate part of the human brain. For the child, the final pathway is determined by the particular language that he or she hears in the surroundings. The mind has been shown to select—and often actively seek out—the kinds of linguistic stimulus it needs to learn language. The infant brain is biologically predisposed to seek out any linguistic stimulus—even nonvocal ones such as sign language. Deaf infants who were born to deaf parents, for instance, were found to "babble" with their hands in bits of sign language—mimicking the language their parents use just as hearing children do.

Over the years, Chomsky and his followers have been mistakenly characterized as contending that all language is innate, that all possible grammars are stored in the genes, that Universal Grammar is exactly identical for all humans, that learning plays no role in the development of language, and that language is unchanging. Such is the range of misconceptions that one linguist in the Chomsky tradition recently wrote an article in a linguistics journal entitled "Why are they saying these things about us?" None of the above views, however, are held by Chomskian linguists. They claim only that there is a fundamental structure that underlies all human languages and arises from an innate property of the human brain.

THE ROOTS OF LANGUAGE

Evidence for the idea that we all harbor an innate language organ in our brain comes from language itself: For instance, people regard Chomsky's famous example *Colorless green ideas sleep furiously* as grammatically correct—even though it doesn't make any sense literally. Conversely, the sentence *furiously sleep ideas green colorless* doesn't seem grammatically correct. The difference in how we view these two sentences suggests that our brain has a mechanism that acts independently of the basic meanings of words to help distinguish the grammatical correctness of sentences. By the same token, there are 3,628,800 ways to rearrange the ten words in a sentence such as *Try to rearrange any or-*

dinary sentence consisting of ten words. Yet only one of the rearrangements of the sentence is grammatically correct and meaningful.

Our uncanny ability to distinguish instantly the one correct sentence from the other 3,628,799 ungrammatical sentences provides strong evidence that the human brain has a kind of mental "cookbook" that contains the basic recipes for producing grammatical sentences. For instance, the sentence *John is easy to please* superficially looks very similar to the sentence *John is eager to please.* But when both sentences are rearranged to *It is easy to please John* and *It is eager to please John,* the similarity vanishes. This is because our brain's book of linguistic recipes knows that at a deeper level the two original sentences are not similar: "John" is the "deep object" of the verb *please* in the first sentence, whereas in the second sentence "John" is the "deep subject" of *please,* and so would have to be transformed to *John is eager to please [somebody].*

Further evidence that language is produced by a specialized mental organ in the brain—and is not merely the by-product of our overall intelligence, as some linguists have proposed—comes from the fact that language is too complex, quirky, and specialized to serve as the mind's main "medium" of thought. The erroneous notion that language reflects the mind's overall thinking ability was proposed earlier this century by Edward Sapir and Benjamin Whorf, who argued that people's thoughts are shaped by the language they speak. Humans "are very much at the mercy of the particular language which has become the medium of expression for their society," wrote Sapir in 1921. "We see and hear and otherwise experience very largely as we do because the language habits of our community predispose certain choices of interpretation."

Though now dismissed by most linguists, Sapir and Whorf's ideas have become a common part of pop culture. Perhaps the most famous example of Sapir and Whorf's notion was the popular assertion that the Eskimos have dozens of words for *snow,* reflecting subtle differences they perceive as a result of their lifelong dealing with the fluffy stuff. Likewise, Sapir and Whorf's theory implied that if a particular language lacked words for specific colors, then speakers of that language would not be able to differentiate between those colors visually. In the same way, some Chinese languages have no explicit form of the subjunctive, and so according to Sapir and Whorf's ideas, someone who spoke Chinese would not be able to think in terms of, *"If I were there, I would . . ."*

But a host of studies suggest that the "language of thought" in the mind is not the same thing as the day-to-day language a person speaks. Anthropologists find that even in particular cultures where people describe colors with only two words—roughly translating to *dark* and *light*—the people in these societies nevertheless have the ability to think about a broad range of colors, categorizing and remembering different hues just as adeptly as people whose language contains numerous color terms. Psychologists have shown that when people are asked to determine whether a drawing depicting two similar geometric figures in fact represents the same figure, one of which has been rotated in space, they perform the task by rotating the figures in their mind—suggesting that they are thinking in purely nonverbal, visual terms. Scores of studies from neurobiology suggest that for the mind to do its thinking in a specific language would detract from much of its enormous power, because the brain operates largely in parallel—that is, over many different cognitive pathways all at once. Language, in contrast, is a singular stream of symbols that come one after another, making it hardly a suitable medium for "language of thought." Lastly, linguists point out that the idea of Eskimos having dozens of different words for snow is a myth: In fact, they have only *two* words for snow: *qanik*, meaning "snow in the air," and *aput*, meaning "snow on the ground."

As most of us have experienced at some point in our lives, language is clearly not the best medium for expressing all the kinds of thoughts we might have. "Language is a way of coding thoughts," says Pinker. "But if you think about a telephone sitting on the table, that thought doesn't have any particular order." What language does, says Pinker, is allow us to translate our murky, fuzzy, infinitely complex thoughts into a discrete set of sounds that can be transferred through time and space to be understood by others. But words and sentences are not thought itself, as those who have struggled to find the right words to express how they feel can testify. Language is a notoriously poor method of expressing emotions, feelings, and the richer tapestries of human existence such as love, faith, beauty, and truth—studies show that people rely more on facial expression and tone of voice to judge a person's emotional state than his or her explicit words. And while language can be used to describe a particular place or scene, for most people a picture really is worth at least a thousand words. If language were perfectly suited to expressing all aspects of human experience, the works of poets, playwrights, and novelists would not be so extraordinary and vital to culture.

In a sense, language is like a "spaceship" for thought, a method by which an idea bubbling within the nurturing, warm confines of the dense interconnections of brain tissue can be distilled, somewhat imperfectly, into a serial string of symbols. This string can be jettisoned into the outside world, where the spaceship of thought can roam freely as speech, a book, or pop song. When it encounters another brain, the symbols are translated back into the fuzzy goop of real thought, complete with the unique associations and permutations of meaning that each person brings to an idea.

FROM THE MOUTHS OF BABES

Further evidence for the innate, biological roots of language comes from how children learn to use proper grammar. Using proper grammar is not just the concern of schoolmasters: By explicitly connecting one set of concepts with others in a particular order, the rules of grammar make possible a level of precision and meaning that would be as vital to a hunter-gatherer as it is to a sophisticated city dweller. Even tiny changes in grammar can make a big difference in the meaning of a sentence. "It makes a big difference whether a far-off region is reached by taking the trail that is in front of the large tree or the trail that the large tree is in front of," note Pinker and Bloom. "It makes a difference whether that region has animals that you can eat or animals that can eat you. It makes a difference whether it has fruit that is ripe or fruit that was ripe or fruit that will be ripe. It makes a difference whether you can get there if you walk for three days or whether you can get there and walk for three days."

Children's ability to quickly learn these complex, subtle grammatical distinctions cannot be merely the result of their using some sort of "all-purpose," learning ability to "solve the puzzle" of how to best communicate with others, as has been suggested by some linguists. "It's sometimes argued that language did not evolve per se, but rather is the by-product of something else that humans are good at, such as learning in general," says Bloom. "But learning a language is not like programming a computer: Parents merely provide children with examples of sentences in English—not the *rules* for how to speak English."

If learning a language simply consisted of figuring out the best way to communicate with others, one might expect language—and language learning—to be very different than it is. "From the very beginning of language learning children conform to various grammatical rules that offer them no clear advantage in communication," says Pinker. "Children will say *big dog*, for instance, placing the adjective before a noun, but they never place an adjective in front of a pronoun, such as *big he*—even though using this convention might give them a broader range of expressibility." Even when the real world provides obvious solutions to communication problems, children apparently don't take advantage of them. For example, a study of how deaf children learn sign language found that deaf children get mixed up just as much as hearing children when expressing the concepts "me" and "you"—even though the hand signs for "me" and "you" consist simply of pointing to oneself or the other person in the conversation—a seemingly far more easy to learn way of expressing the concepts. Furthermore, what might seem an obvious solution to solving the problem of communication in one person's opinion may not be so obvious to some other person. "Even iconicity and onomatopoeia are in the eye and ear of the beholder," say Pinker and Bloom. "In American Sign Language, the sign for 'tree' resembles the motion of a tree waving in the wind, but in Chinese Sign Language it is the motion of sketching the trunk. In the United States, pigs go *oink;* in Japan, they go *boo-boo.*"

Another reason that it would be nearly impossible for a child to learn language simply by generalizing from a set of examples is that language is full of red herrings: Hearing the sentence *John saw Mary with her best friend's husband,* followed by the sentence *Who did John see Mary with?,* one might conclude that the sentence *John saw Mary and her best friend's husband* could be turned into the ungrammatical *Who did John see Mary and?* Likewise, hearing *The baby seems to be asleep* and *The baby seems asleep* might suggest that *The baby seems to be sleeping* can be truncated into *The baby seems sleeping.* Although many of the examples of language that children must learn from seem contradictory, children don't appear to go through a period of extensive trial and error learning a language, as one might expect if they were picking up the fundamental structure of language merely from generalizing from the speech they hear from others. Children sometimes make minor grammatical errors, but they never make whopping mistakes such as saying *Milk some want I.* "These language puzzles may seem kind of silly," says Bloom. "But you can't help but be impressed with the rich complexity of things. If you look at the speech of a three-year-old, you are immediately struck by something: It's no accident that a computer can't produce or comprehend speech. It is not because we need to build fancier computers—it's that we simply do not know how even a three-year-old does it. We know the rules for playing chess, we can program the rules of baseball into a computer, but we are miles away from doing language, because it is so astonishingly complex."

Despite the enormous complexity involved in learning a language, children rarely are "instructed" by their parents on whether the sentences they speak are grammatical. If a child says, "Nobody don't like me," for instance, a parent is just as likely to reply, "Of course they do!" Even when parents do try to correct their children, the kids often steadfastly stick to their conceptions of how the wording should be. In one exchange monitored by linguists, a child repeated, "Nobody don't like me" ten times, with his mother correcting him—"No, say, 'Nobody likes me' "—each time. Finally, the child exclaims in a moment of epiphany, "Oh! Nobody don't *likes* me." And while children rely on the samples of language they hear to learn the vocabulary and superficial grammatical distinctions that make English different from Japanese, for instance, it is clear that their knowledge of language extends far beyond those specific examples. A child might hear a grammatically correct sentence such as *John must have been being tickled,* says Pinker. But they also know that *John must have being tickled* is "word salad." "Certainly no one *teaches* you that," he says.

Children seem quite adept at learning language even when there is little language in the environment to imitate. In one study of how children learn language, linguists examined the communication skills of severely deaf children who were raised in a family where the parents knew no formal sign language. Despite the fact that the parents used simplistic hand gestures in an unstructured way, the deaf children nevertheless developed a languagelike signing system of their own, with a specific sign-ordering syntax and prepositional phrases. A similar phenomenon took place during colonial times, when speakers of many different languages were thrust together on isolated islands, typically because of slavery. With no native tongue in common, these various peoples spoke a form of speech known as *pidgin*, which relied on very basic combinations of simple words and phrases with none of the richness of human language. Within a single generation, however, the children in these communities used this nonlinguistic input to invent their own complex, highly structured languages, which are called *creoles*.

Further evidence for the brain's innate ability to seek out the information it needs to acquire a language comes from the fact that there is a critical period in which children are primed to acquire a language. This period, beginning around the eighteenth month of a child's life and lasting until about seven years of age, also corresponds to a surge in the growth of neural circuitry in the brain. In one study, linguists examined the language abilities of more than fifty deaf people who learned American Sign Language at various times in their lives. The researchers found that those people who had been exposed to ASL between birth and about age seven made fewer errors than those who learned ASL later in life. The researchers found similar results with non-English-speaking immigrants who came to the United States at various ages and then learned English. Those who came to the United States before the age of seven were more adept at recognizing ungrammatical sentences than those who learned English later in life.

While children are undoubtedly influenced by their surroundings as they learn a language, they nevertheless follow a program that seems more akin to learning to walk than learning to play the piano. That is, despite minor variations, all children the world over learn language more or less at the same time in more or less the same way, even when the children receive no explicit instructions. By the age of six months, infants typically have moved from cooing to babbling, and can distinguish between speech sounds such as a *P* or *B*, or a *T* or *D*. By the time they've reached eight months their babbling has taken on a semblance of the speech sounds of their environment, to the extent that an English-speaking mother, for instance, can pick out an infant raised in an English-speaking household from infants who were raised in homes where the native tongue is Russian or Chinese. Most children begin speaking single words at about their first year—the exact timing of their language development has no relation to their adult language abilities or their overall intelligence. At eighteen months children begin speaking in what is called *telegraphic* speech: strings of two to five words such as "more juice" and "I carry" that typically omit prepositions and other complexities but are rich in meaning, as every parent knows. By two and a half, children typically are capable of speaking in complete sentences. The acquisition of specific parts of language seems to follow a distinct pattern common to all children, regardless of the economic or social background. English-speaking children learn the *-ing* form of a word, for instance, as in *doggie barking* before they learn to combine it with the verb *to be*, as in *doggie is barking*. They learn to use the prepositions *on* and *in* before any others. And they learn to use *a* and *the*, then the possessive *'s*, then contractions such as *isn't* and *don't*.

As with all of our evolved mental "organs," the biological underpinnings of language require some environmental input if they are going to be expressed. The importance of the environment in learning language was tragically demonstrated in the case of a girl known as Genie. From the age of eighteen months until age thirteen Genie was imprisoned in her bedroom alone by her parents and was not exposed to any language. When she was finally discovered by Los Angeles authorities, she was unable to speak. Possessed of a normal intelligence, Genie was able to learn to speak in short, spare bursts of words after many years of care and training. But she never got beyond the verbal abilities of a two-year-old.

The case of Genie provides a further, graphic example that the supposed dichotomy between nature and nurture is a meaningless one. As in many aspects of human behavior, genes and environment are not separate, conflicting influences on the development of the human mind. Rather, they work in concert, and are designed to be that way: The need for environmental input is a fundamental aspect of a human's genetic programming. Children's surrounding environment is not only a vital component of their language development, it is vital to the workings of language itself. While the fundamental grammatical structure of a language may be part of the evolutionary legacy of the mind, much of a particular language—its superficial grammar, vocabulary, idioms, etc.—are "stored" in the environment. That is, these aspects of language are maintained by the at-large community culture in which a child grows up.

TALKING PRIMATES

Researchers have long looked for clues to the evolution of human language by studying the language abilities of our close evolutionary cousins. Yet, ironically, while these animal studies have succeeded in blurring the line that has been assumed to exist between human language and the communication skills of other animals, the biggest effect of these investigations is to reveal just how sophisticated human language is. Although research demonstrates that some animals can learn the rudiments of vocabulary and word order, it also shows that the human capacity for language surpasses either of these abilities.

For instance, vervet monkeys have as many as six different alarm calls for the various predators that feed on them. These vervet calls are not merely general calls of alarm: One call indicates the presence of a leopard or other catlike predator, for instance, causing the monkeys within earshot to scramble up into the safety of the trees. Other calls elicit different types of behavior: A gruntlike call warning of the presence of eagles causes vervets to look up into the air, and a different call causes them to look down in search of pythons; there are also calls for hyenas, baboons, and unfamiliar humans. Most of the vervets who hear the calls exhibit the correct response typical for each predator, even though there is no predator in sight, suggesting that sound alone is enough to evoke the proper reaction. Rather than being simply an uncontrollable vocal reaction to the sight of a leopard or hawk, for example, the calls appear to be voluntary: Vervets do not call when they are alone, for instance, and call more frequently when relatives are present versus nonrelatives.

While the various calls used by vervets might appear to be a form of language, there is a subtle distinction between human and vervet communication. In humans a simple word such as *python*, for instance, can have many different interpretations. Said with a rising intonation at the end of the word would be asking, "Is there a python over there?" Whereas said with a neutral tone, the word might merely be a response to a question, such as, "What's the biggest African snake?" Moreover, these uses of the word *python* do not require that the speaker be referring to any one snake in particular: In human language, words do not directly refer to things that exist in the world but are used to describe our mental *representations* of things. That is, the information coming into our senses about a python is gathered together to form a mental representation of the snake, and it is this mental representation, not the snake itself, that is used as a basis for the word. This mental "translation" makes possible words such as *unicorn* and *Valhalla*, which refer to nothing at all in the real world. In contrast, the vervet's python warning call has only one meaning: A python is present and poses a danger.

HAVING THE RIGHT INTENTIONS

A critical difference between vervet calls and human language involves the speaker's and listener's ability to form a theory of mind about each other. Philosopher Dan Dennet of Tufts University, who has accompanied primatologists Robert Seyfarth and Dorothy Cheney on their expeditions to study vervets in Kenya, argues that a creature's capacity for forming a theory of mind can be ranked according to what he calls *levels of intentionality*. These levels correspond to the amount of knowledge a creature has about its own mind and the minds of others. For example, if vervets had minds that merely display what Dennet characterizes as "zero order" intentionality, their calls would be an unthinking reaction to the sight of a predator, much the same way that a sharp pain might make someone yell "OUCH!" If vervet minds displayed "first order" intentionality, their calls would be roughly equivalent to saying, "Quick, climb up into a tree!" That is, the sight of a predator would cause a change in the vervet's own mind, but the intent of its call would not be to change the minds of other vervets—the vervet would not be cognizant that the minds of other monkey's might be different from its own. The intent of the call would be to alter other vervets' *behavior*, by telling them to climb a tree.

As Dennet's levels of intentionality go upward, the layers of understanding of others' states of mind multiply. If vervets had minds of "second order" intentionality, for instance, their calls would roughly translate to, "Hey, you should know that there is a python nearby"; that is, the caller recognizes that the minds of other vervets might be different from its own, and the call is an effort not just to change other vervets' behavior but to change their minds as well. In a "third order" intentional system, the caller not only is aware that the minds of other vervets may be different from its own mind but it also knows that these vervets know that its mind is different from theirs. Its calls are an attempt to make other vervets aware of a change in the caller's *own* mind—as in, "Hey everybody, I see a python and I think you should head for the trees!"

Human language typically involves at least a third-order intentionality: Dick talks to Phyllis because he wants to change her mind about what is going on in *his* mind. In other words, he talks because he wants to be understood. Because the human brain has been crafted by evolution to be superbly adept at predicting the minds of others, the real meaning of someone's statement may lie beyond the actual text of the communication. "Humans are intentional systems almost to a fault," note Cheney and Seyfarth. "Our elections reveal with depressing consistency that politicians worry more about what others think (and what others think they think) than the actual course of events."

Humans are capable of dealing on an infinite number of levels of intentionality, notes Dennet, who wryly adds: "But in fact I suspect that you wonder whether I realize how hard it is for you to be sure that you understand whether I mean to be saying that you can recognize that I can believe you want me to explain that most of us can keep track of only about five or six orders, under the best of circumstances." Happily, such convoluted levels of understanding (and misunderstanding) are rare. But our amazing ability to appreciate the various levels of each other's mind is what adds layer upon layer of complexity to human relations: We appreciate a compliment all the more for our knowledge of how hard it is for the complimenter to make such statements, for instance. Likewise, a person might be hurt if someone interprets as an insult a statement that was intended to be perfectly harmless, scolding that "I thought you knew me better than that."

The calls of vervets fall short of humanlike language, because the monkeys appear to display only "first order" intentionality. That is, the purpose of their calls is to change others' behavior, but not their state of mind. Evidence for this conclusion comes from the fact that, occasionally, a low-ranking male vervet will give a false call indicating the presence of a leopard, for instance, when he sees a rival male approaching the group. This bit of deception typically causes the rival to scurry up a tree, aborting his attempt to join the group. While the deceiving vervet clearly wants his rival to climb into the trees, however, he appears *not* to be trying to change his rival's mind. Often the deceiving vervet will simply stroll along the ground issuing "leopard" warning calls, apparently oblivious to the fact that his own behavior shows no signs that the alleged leopard is actually present. This is a bit like sitting calmly in one's chair and yelling "Fire! Fire!" to the other people around you in the room. The fact that the deceiving vervet makes no attempt to behave as if a leopard is nearby suggests that the monkey is merely taking advantage of the tendency of other vervets to scamper into trees when they hear a warning call—not that it is trying to actually change the other vervet's mind about whether a leopard is there. In other words, the vervet utterances lack the complexity of human language.

SIGNS OF LANGUAGE?

Another avenue of exploration in the search for the origins of human language is the numerous attempts to teach apes to communicate using American Sign Language or a set of symbols. Perhaps the most famous of animal signers, the chimp Washoe, learned hundreds of the signs that are used in American Sign Language. Washoe used signs to denote classes of objects rather than individual objects themselves—knowing that the sign for banana, for instance, refers to *any* banana, not just the one in front of her. Washoe also applied words she had learned to new objects: She called a nightcap a hat, even though she had never seen that kind of hat. The first time she saw a duck, Washoe called it a "water bird." Apes also recognize that word order conveys meaning. Using an array of plastic symbols mounted on a board, a bonobo chimpanzee known as Kanzi spontaneously began to use word order as part of the meaning of what was said. The order in which Kanzi signaled two words such as *chase* and *hide*, for instance, reflected the order in which Kanzi carried them out. The ability to use word order to convey meaning has also been demonstrated in other large-brained mammals, such as dolphins and sea lions.

While impressive, these studies ultimately reveal the wide chasm between human and animal communication. For instance, the most signs claimed to have been learned by an ape is about 400—they typically learn far fewer—whereas a child will know as many as 10,000 words by age ten. And while word order is an important part of language, there is more to understanding what is being said than the order in which the words appear. There is a crucial distinction between *John shot Mary* and *John was shot by Mary*.

Ultimately, these animal language studies reveal what seems to be an unspoken assumption among researchers that human language is some sort of pinnacle of biological communication—something that other animals would use if they could, but can only aspire to. "A student once asked me whether, if we wait long enough, chimpanzees will someday evolve language," says Bloom. "It's as if he thought that human language is truly where apes want and ought to be. But this idea that human language represents the pinnacle of evolution is really a silly game to play: You could imagine a dung beetle bragging about how many balls of dung he can roll." Human language is remarkable, yet our preoccupation with whether animals can mimic human language masks the fact that animals are extremely adept at communicating *without* using humanlike language. "The question of whether animals have *true* language is meaningless," says Pinker. "It's like asking if animals have 'true' vision."

Our propensity to focus on whether animals display humanlike mental activity—while ignoring an animal's other cognitive powers is aptly demonstrated by the story of "Clever Hans." At the turn of the century, Wilhelm von Osten, a retired school teacher in Germany, caused a sensation with his horse, Clever Hans, who had apparently learned how to solve mathematical problems. Asked to subtract seven from twelve, for instance, the horse would tap five times with his hooves, and could do it even when von Osten was absent from the room. Eventually, however, it was revealed that Clever Hans knew more about human psychology than mathematics: The horse's apparent skills in math arose from his ability to read subtle cues from the faces of his flabbergasted audience. He merely kept tapping until their faces registered that he had reached the correct number. People had been fooled, including von Osten, because they were unaware that they were communicating with the horse nonverbally.

Critics of ape language studies contend that the apes, too, are simply responding to nonverbal cues from their trainers that they are giving the expected hand signals. But even when steps are taken to eliminate nonverbal cueing, such as having Kanzi use a keyboard of symbols, it is not clear that the animals are doing anything but putting together strings of symbols to get food and other rewards from their trainers. In humans, on the other hand, language reflects not only our mundane needs but also our ethereal desires. Even two-year-olds use language to share with others a host of seemingly irrelevant information about how they observe the world, themselves, and their inner thoughts and feelings. It is precisely these other functions—many of them social, not informational—that may form the bulk of the evolutionary significance of language for humans.

It may be that human language is "special" like the giraffe's long neck or the bat's ability to use echoes to navigate—different, but not necessarily better. You could argue that language is necessary for human survival precisely because humans lack those other "special" apelike adaptations that enable other primates to survive without language. Ultimately, teaching apes to use sign language may be a bit like teaching humans to swim; the ability of one species to mimic another species' adaptations reveals little about either. Our preoccupation with the linguistic abilities of animals is like our being overly awed by Clever Hans's ability to add and subtract. Humans seemed more impressed by the horse's ability to perform a computationally trivial feat such as adding two numbers, which can be performed by an *extremely* dumb calculator, rather than Clever Hans's far more sophisticated ability to detect and understand the subtle clues in people's faces—something the world's most powerful computers cannot come close to doing. Being able to judge the thoughts of other beings without language is a more important aspect of intelligence than knowing how to add two and two—or perhaps even learning language.

WHAT GOOD IS HALF A LANGUAGE?

The evidence that language is an evolved mental organ still leaves open the question of exactly how language evolved. Like the human eye, language depends on the complex interaction of many specific structures, such as tense, subject, object, possession, etc., and so it is hard to imagine a language that only had a few of these features. "Some people might ask, 'What good is half a language?' " says Bloom. "And it's true that if you ripped it in half, it wouldn't work. It's a complicated machine, and a complicated machine won't work if half of it is gone. But the real question is not whether a half a language can work, but whether you can build it up piece by piece. Half an eye may be useless, for instance, but a *rudimentary* eye is not useless. The same thing is true of a rudimentary language." A rudimentary language system that lacked the complex structure of modern human language nevertheless might still be useful. Such "intermediate" types of language exist today: As Pinker and Bloom note, "pidgins, contact languages, Basic English, and the language of children, immigrants, tourists, aphasics, telegrams, and headlines," all demonstrate that language can operate at a wide variety of sophistication and simplicity and still be quite useful for communication.

A subtle way that human language ability could have gradually become more complex over time is by evolving through a mixture of nature and nurture. As a simple example, suppose that there was a sophisticated human ability—such as language—that depended on a complex interaction of some twenty separate traits. Suppose further that all of the traits are necessary for the behavior to be beneficial, that is, those people born with only some of the traits would have no advantage over others born with none of the traits. The chances against this complex ability arising through a sudden burst of simultaneous genetic mutations that produce all twenty necessary traits is astronomically large. What's more, even if a person were born with the gene for all twenty traits, this genetic arrangement would be lost in the next generation, because sexual reproduction splits up the parents' genes.

But now suppose that these traits could be learned through trial and error, even if they are not encoded in a person's genes. Paradoxically, the fact that these traits could be learned would lead to greater evolutionary pressure to make the traits more innate. The reasoning is as follows: While the chances are remote that a person might get the genes for all twenty traits at the same time, the odds are greater that a genetic mutation could result in a person being born with *10 percent*, say, of the necessary traits encoded in their genes. Those individuals with 10 percent of the traits innately present gain an evolutionary advantage over those people who had none of the traits in their genes, because they would need to learn fewer traits, and so would acquire the ability faster. If the offspring of these people had another mutation that gave rise to another 10 percent increase in innate elements, these people with 20 percent of the necessary traits in their genes would learn the ability even faster. If their offspring had a mutation that increased the innate component of the ability another 10 percent, the advantage would be greater still, and so on, until most of the traits necessary for the ability were fixed in the genes. Interestingly, as more of the traits became innate, the evolutionary advantages to having even more traits being innate might slowly decrease. People with 95 percent of the traits in their genes, for instance, would have little selective advantage over those who had only 90 percent of the traits in their genes, because the bulk of the traits would be present and everyone could soon learn the rest of the ability.

While the evolution of language was no doubt more complex than this simple example, it suggests an intriguing explanation for the combination of nature and nurture that characterizes human language today. One way to look at it is that the fundamental structure of language lies in the genes, because language's infinite variations of words and phrases make it impossible to have all possible examples of grammatically correct sentences in the environment. On the other hand, the vocabulary and superficial grammar of a language can be "stored" in the surrounding social environment of everyone else's conversation, and so can be learned. This is advantageous because it allows words to change along with the external environment with the introduction of new objects (such as lasers) or new circumstances (such as jetlag).

From the innate mental machinery that makes a child an expert, by age three, of one of the most complex features of human culture to the deep connections that language forms among people around the world, language is vivid testimony that humans are more alike than they are different, and more of an evolutionary family than we might imagine. That this quintessential behavior has ancient, common roots is a fitting—and living—demonstration that all the inventiveness of human culture cannot mask the deep connections between members of our species no matter where they live. These deep-seated connections received further bolstering by new research into the dawn of the human species, which is shedding light on perhaps the most fundamental step in the evolution of the psyche: the beginning of culture.

Chapter Eight

―――――――

THE

CREATIVE

EXPLOSION

Culture may even be described simply as that which makes life worth living.
—T. S. ELIOT

About 40,000 years ago, the evolved psychology of our ancient ancestors blossomed in a way that had never before been seen on the face of the earth. After eons of displaying the simplest of cultural behaviors, people suddenly embarked on a radically new way of making their living: They crafted new types of elaborate tools from new kinds of materials such as bone and ivory. Their lives were infused with music, dance, and ritual, including elaborate burials of their dead. They lived in new kinds of sturdy houses, warmed themselves with newly designed high-tech fire hearths, and tailored exquisite clothing to protect themselves from the Ice Age winters that gripped the Northern Hemisphere. Their dinner menus featured a bounty garnered with a leap in sophistication in their skills in hunting and gathering. For the first time ever, they made art: They crafted fox teeth and stone into amulets that dangled from necklaces, carved vivid depictions of animals into the handles of their tools, sculpted tiny, "portable" statuettes of leaping horses and other animals, created enigmatic figurines that suggest the religious worship of a goddess, and sewed hundreds of tiny beads into headbands, shirts, belts, and other clothing. And in perhaps the most famous examples of Ice Age art known today, they painted glorious images of deer, bison,

and other animals on hundreds of caves in France and Spain.
The creative explosion had begun.

Yet, far more than mere tools, clothing, and art, the biggest innovation in this mysterious explosion of culture forty millennia ago was *innovation* itself. Before that time, the simple stone tools made by our ancient human ancestors remained monotonously unchanged in their design for literally tens of thousands of years throughout the world. But this new chapter of the human saga celebrated change for change's sake, as tool designs differed from year to year and from valley to valley. "There was more innovation in the first five minutes of this period than in all the human history leading up to it," says archaeologist Randall White of New York University.

White is one of a new breed of archaeologists who are using the artifacts left behind by our ancient ancestors to trace the evolution of the human psyche—particularly its social evolution. While much of the attention on paleoanthropological research over the years has been focused on finding the very oldest human ancestors, White is interested in the more recent, final stage of our evolutionary saga: the origins of our own species, *homo sapiens*. He is particularly interested in the abrupt, crucial shift that occurred some 40,000 years ago, when our ancient ancestors began *acting* like modern humans, displaying the kinds of cultural behaviors we exhibit today.

White's research is concentrated on a vital aspect of everyday life that appeared at nearly the beginning of the creative explosion: the dawn of art. At his NYU laboratory in Manhattan, he excitedly opens drawer after drawer showing dozens of carved statuettes, beads, pendants, and pierced animal teeth dating back as far as 34,000 years ago—which is tens of thousands of years before the majestic cave paintings were made. These objects signal the beginnings of our self-awareness as a species: White holds his hand up to the light, his thumb and forefinger squeezing an ancient, tiny rod of ivory smaller than a pencil. The rod is topped with a minuscule engraving of a bearded man who appears to be wearing a cloak—one of the oldest images of the human form that exist.

These pieces of art reveal that the creative explosion that took place some 40,000 years ago was as much a social revolution as a technological one. Suddenly, *style* became a very important part of our ancestors' everyday lives: The kinds of objects and ornamentations people wore became an important indicator of their affiliations and position in the group at large, and took on symbolic meanings as to who they were and what they cared about. It is a practice that should seem familiar to us modern folk as well, for whom cars, jewelry, T-shirts, and caps often broadcast our status, ideology, personal taste, and view of the world. "There is not much difference between the kinds of body ornamentation we see in the creative explosion 40,000 years ago and a modern-day person's school tie, or the insignia on the back of a motorcycle jacket," says White. "It's group identity, it's solidarity. It's human culture as we know it—a pattern of behavior that, if you saw some people in some corner of the modern world behaving this way, it wouldn't seem outlandish. But back *before* 40,000 years ago, none of this exists. It's absent."

THE HUMAN REVOLUTION

The search for the origins of this dramatic shift in our species' psychology begins with one of the most controversial questions in paleoanthropology: When, where, and how did our own species, *Homo sapiens*, arise on the planet? For many years, most researchers believed that the various races of modern humans around the world arose from the populations of the ancient human ancestor known as *H. erectus*, which had moved out of Africa about a million years ago and migrated into Europe, Asia, and the Middle East. According to this view of our past, which has been dubbed the "multiregional" model of human origins, the various African, Asian, and European races of modern *Homo sapiens* have very ancient roots: Modern Asians, for instance, are thought to be the descendants of the population of *H. erectus* that lived in Asia hundreds of thousands of years ago. Likewise, this multiregional scenario regards the Neanderthals, the brainy, hulking, beetle-browed hominids who lived in Europe for more than 100,000 years up until about 30,000 years ago, as the direct ancestors of modern Europeans.

More recently, however, a growing number of researchers are arguing that, in fact, only a tiny number of the world's population of *H. erectus* were the ancestors of modern humans. In this scenario, humans are thought to have arisen from a small population of *H. erectus* who lived in Africa some 200,000 years ago. Some of these modern humans then migrated out of Africa and swept into the rest of the world—eventually displacing all the other somewhat brainy, two-legged, humanlike creatures who were roaming the earth at the time. In this "Out of Africa" scenario, the "creative explosion" in Europe marked the arrival of these modern humans—who are known as the Cro-Magnon, after the French site where their bones were first discovered—into a part of the world that previously had been occupied solely by Neanderthals. Eventually, the Neanderthals became extinct.

The replacement of the Neanderthals by modern humans need not conjure up images of bloody clashes between the two types of hominids or wholesale genocide. A computer simulation of the interacting populations reveals that if the Neanderthals had as little as a 2 percent lower birth rate as a result of the influx of modern humans, they would have gradually disappeared in a couple of thousand years. As it is, the latest archaeological evidence suggests the modern humans and Neanderthals may have coexisted in Europe for as long as 10,000 years. "It's an awesome thought that we need go back only 40,000 years to have more than one species of humanity on the earth," says White. "There is this fantastic moment in human evolution, where Neanderthals and modern humans are living virtually cheek by jowl."

ALL ABOUT EVE

Originally, the "Out of Africa" model was based only on fossil evidence. But the scenario got a boost by startling new findings from geneticists. Comparing snippets of genes taken from people around the world, researchers found evidence that all humans alive today share as a common ancestor a woman who lived in Africa some 200,000 years ago.

Inevitably, perhaps, this woman has been dubbed "Eve." But in fact she is ill-named, because she is not the mother of all humans. Eve's connection to the modern human race exists through a quirky bit of genetic material known as mitochondrial DNA. This mDNA, as it is called, resides outside a cell's nucleus—the place where the ordinary DNA that serves as the blueprint for the entire body exists—and instead lies

within tiny organlike structures within the cell that are called mitochondria. Compared to the rest of the human genome, mDNA is a tiny,
insignificant sliver of a molecule, making up only 1/3,000 of the DNA
in a human.

But mDNA does have one unique feature that makes it most important to geneticists, and, ultimately, to researchers studying the origins of
the human psyche: mDNA doesn't take part in the genetic reshuffling
that occurs every time a child is conceived. When conception takes place,
half the father's DNA and half the mother's DNA combine to make a
new person with a newly reshuffled set of genes—which is why people
tend to look like both their parents. At the same time, however, a fetus
begins its development in the womb as a single cell, the mother's fertil-
ized egg. Since the egg is made by the mother, it is full of copies of the
mother's mitochondria and their intact, unshuffled mDNA. This egg cell
divides and replicates again and again to become the fetus, with the result that virtually *all* of a person's mDNA comes from his or her mother.
From a woman to her daughter to her granddaughter to her great-grand-
daughter, and so on, the mitochondrial DNA in each woman stays the
same through the generations.

Over time, however, the tiny sliver of mDNA does change slightly
through random mutations, typically from the mDNA molecule making
mistakes in copying itself as the cell divides. This sometimes produces
slight differences in the mitochondrial DNA from one generation to the
next. While a mother and daughter's mitochondrial DNA might differ
only slightly, the differences between the mitochondrial DNA of a
mother and her great-great-great-granddaughter would be greater, because more time has elapsed between those many generations—time
during which mutations might have occurred during copying.

Geneticists realized that if the rate at which mDNA mutated was constant and could be estimated, mDNA could be used as a molecular
"clock" to trace the human lineage back to its beginnings. In some kinds
of molecules, mutations take place on a fairly regular schedule: Over a
million years, for instance, 1 percent of a particular molecule might undergo mutation. If scientists could measure the amount of variation between various samples of mDNA, and could estimate at what rate the

mDNA mutated, they could estimate how long ago those particular samples were first separated. That is, researchers could trace the pieces of mDNA in the world's humans back to their common, ancestral origins. The geneticists compared samples of mitochondrial DNA from people all over the world. The results of their study were breathtaking: They found that the mitochondrial DNA of humans today is remarkably similar, suggesting that our species has very recently sprung from a common source. According to the mutation rate for mDNA estimated by the genetic researchers, all humans alive today share a common ancestor who lived only about 200,000–100,000 years ago.

Unlike the biblical Eve, this common ancestor by no means gave rise to all the people who ever lived. Rather, this woman is simply the only person whose descendants have had a daughter in every generation up to the present. It is impossible, in fact, that somewhere along the human lineage such a woman could *not* exist, because for that to happen, at some point the entire population of women on Earth would have to have had nothing but sons. As an analogy, consider how family names are passed on over generations in cultures where the male's surname goes to a couple's children: If parents have only daughters, a surname might be lost forever—even though the genes of both parents are still carried on into the next generation. Thus if the world consisted of nothing but people whose last name was Jones, it would not mean that someone named Jones was the ultimate father of all people on Earth. It would mean merely that Jones's descendants were the only people who had at least one son in every generation. Just as a last name will disappear, so, too, have strands of mitochondrial DNA died out as no female offspring appeared in a particular generation.

The Eve ancestor can't be the sole source of all human DNA, because her legacy of nuclear DNA—the stuff that largely determines a person's whole being—was cut in half at once when it was mixed with a male's genes to produce Eve's children. Consider the fact that you have two parents, four grandparents, eight great-grandparents, and sixteen great-great-grandparents: Your genes are a mixture of the genes of all sixteen of those great-great-grandparents. Yet your body's *entire* collection of mDNA came from only one of these sixteen ancestors: your mother's mother's mother. The mitochondrial Eve, then, can be thought of as merely one of humankind's 10,000th-great-grandmothers.

As might be imagined, the "Out of Africa" model is extremely controversial. Mathematical studies show, for instance, that it is hard to pin down conclusively from the gene studies that the Eve ancestor lived in Africa and not somewhere else, like Asia. But the theory is supported by archaeological and fossil evidence as well: The oldest modern human fossils and human artifacts, for instance, dating back some 100,000 years, are found in Africa.

Whether the source of modern humans turns out to be Africa, Asia, or whatever, the notion that the modern human form has only recently sprung up from a common root has profound implications for understanding the origins of the human psyche. The theory implies, for instance, that all humans are part of a very closely knit biological family. The world's humans have only one-fortieth to one-fiftieth as much genetic diversity as a whole, for instance, as our closest evolutionary cousins, the chimpanzees. Furthermore, the "Out of Africa" model suggests that the racial differences we see among peoples today are very recent in origin and superficial, compared with our species' overall body structure and, most important, psychology. In fact, there is more genetic variability among members of any particular race than there is between the different races as a whole, meaning that it is possible for a white person, for instance, to be genetically more similar to an Asian or African than to another white. We all share a fundamental, universal human blueprint for biology and psychology that transcends the superficial manifestations of a person's color and culture.

Most important, the "Out of Africa" model implies that the human species is not some sort of *inevitable* result of evolution. The multiregional model suggests that the various lineages of ancient hominids the world over gradually took on the modern human form, as if who we are today represents some sort of natural point of convergence for all these various humanlike creatures. The "Out of Africa" model, on the other hand, is more in keeping with the fact that evolution has no overall agenda or ultimate goal: Modern humans are merely one highly successful example of the many different kinds of brainy, two-legged primates that peopled the earth over the past 3 million years. The emergence of modern humans was an isolated, unique event—but no more unique, predestined or predictable than the appearance of any other species.

THE CREATIVE EXPLOSION

The fact that "anatomically modern" humans existed some 200,000–100,000 years ago—at least 60,000 years *before* the appearance of art, jewelry, and sophisticated technology among our ancestors—poses what is perhaps the deepest mystery in the search for the origins of the human psyche. If the human form has been in existence for more than 100,000 years, what was it that caused this creative explosion to happen much later? "Beginning 40,000 years ago, you have people who suddenly latch on to these very complex ways of doing things that didn't exist 500 or even a 1,000 years previously," says White. "Suddenly, something happens, in which you get this new spear point technology, you get people decorating their bodies, you get people sculpting little animals out of ivory. There is a whole constellation of behaviors that suddenly show up that are completely different from anything we've seen before." Fossils of anatomically modern humans who lived in the Middle East as recently as 45,000 years ago are found along with the kinds of simpler tools that are typically associated with the Neanderthals. Likewise, there are no signs that anatomically modern people engaged in "behaviorally modern" styles of hunting, constructing elaborate dwellings, or making art prior to about 40,000 years ago. "In the period between 100,000 and 40,000 years ago, we have anatomically modern humans—we have *us*," says White. "But where's all the stuff? Where are the bone spear points? Where's all the body ornamentation? Where is all the art?"

The change in behavior is so abrupt that some researchers have suggested that our species must have undergone some kind of profound genetic mutation 40,000 years ago—which perhaps, some argue, suddenly gave rise to our elaborate language abilities. Yet considering the human psyche, this seems unlikely: Art and music also played a crucial role in the creative explosion, and musical ability and artistic prowess are fundamental aspects of the brain that are unrelated to linguistic abilities.

People who suffer brain damage through strokes, for instance, sometimes lose their musical talents without losing their language skills; in other cases, victims of brain damage lose their linguistic abilities but do not lose the ability to engage in other cultural behaviors such as dancing, playing music, or creating new tools. Also, fossil evidence suggests that the capacity for language probably existed in some form, albeit rudimentary, among the Neanderthals and more ancient human ancestors as well, and these people did not display the rich creative powers of the Cro-Magnon. It is more likely that the ability to communicate in complex language was not the cause of the creative explosion but one of many *consequences* of some other, more fundamental change in human behavior that made new kinds of tools, art, and music more useful to people and their society.

Fossil skeletons suggest that our ancestors may have begun this fundamental change in their psychology long before the appearance of the cultural explosion itself. Hominids such as *H. erectus* and the Neanderthals were large, heavily muscled creatures whose thick, battered fossil skeletons bear witness to the enormous wear and tear of a life that depended a great deal on sheer physical exertion to survive. The patterns of stress on fossilized Neanderthal skeletons suggest, for example, that adult males could have lifted as much as 700 pounds. In contrast, the fossils of even the earliest anatomically modern humans are more lightly built than their predecessors, and their bones display less evidence of everyday wear and tear. This less rugged physique of modern humans could not have arisen by accident: Though these ancient humans still lacked the cultural signs that marked the creative explosion, they clearly were doing *something* differently in their everyday behavior, because they were able to survive without encountering the enormous physical stress experienced by their forebears. "The question is, what kind of advantage did anatomically modern humans have when they appeared?" says White. "What was so special about this little creep with a big skull—who was basically the nerd on the beach, who all the other guys were kicking sand at?"

• • •

THE SOCIAL REVOLUTION

The revolution in our ancestors' psychology was not simply the ability to make new tools or to use language, but to create a dramatic leap in the complexity of human society itself. "There almost isn't a terminology to describe what happened," says White. "It was a change in social organization, a whole new social matrix. And as a result of these social changes, one cultural group becomes dominant and expands outward, and comes to encompass the entire globe." An analogous "explosion" of behavioral change occurred some 10,000 years ago, when humans around the world began farming. Within 5,000 years after the dawn of agriculture, the entire planet was more or less populated with farming societies that displayed a very different social structure and culture than their hunting and gathering forebears. If future archaeologists looked back at that time through the lens of 50,000 years, they would see an apparently abrupt shift, too. "Some researchers like to see the creative explosion as coming as a result of some sort of biological change, because it is a quick and dirty and efficient way to resolve the problem," says White. "But the differences in human behavior before and after the dawn of agriculture are probably as great as or greater than those in the creative explosion—and no one would *dare* suggest that people who lived 10,000 years ago were somehow neurologically different from people before them."

The social organization of the Cro-Magnon was very different from that of the Neanderthal: The Cro-Magnon's personal ornaments testify that each individual's identity was important to each other—and to the group at large—and the elaborate burials of some Cro-Magnon individuals indicate that these people held a certain status among a group, suggesting a complex social structure. In contrast, while the Neanderthals appear to have also occasionally buried their dead, the burials were fairly simple affairs—earlier suggestions of elaborate burial rites among the Neanderthal are now believed to have been mistakenly exaggerated—and the Neanderthal burials may have been the result of a simple urge to dispose of dead bodies. The Neanderthals left behind no art, personal or otherwise, nor did they display the kinds of hunting prowess that would have come from a highly cooperative social structure.

The complexity of the Cro-Magnon social structure is also reflected in the "oddities" of modern human biology. Humans are unique among primates in that infants are quite helpless when born and require a good deal of care and instruction for more than a decade, reflecting the need to have a dense social structure that would provide care. In contrast, Neanderthal children appear to have been quite robust very early in life—studies of one fossil of a three-year-old Neanderthal suggest that it could lift some 200 pounds—indicating that like many primates, Neanderthal children matured early.

In addition, modern humans have a very long life span compared to other primates. The potential for humans to live a long time is not merely due to improved nutrition and medicine: In societies around the world, at least a few people live into their hundreds, and a large number of individuals live to be seventy or eighty—no matter what kind of technology the society has. Skeletal evidence reveals that some ancient Cro-Magnon lived well into their fifties, too. In contrast, studies of adult Neanderthal fossils indicate that most died in their early thirties. This short life span, coupled with their rapid juvenile development, suggests that the Neanderthals had life cycles similar to other primates such as chimpanzees.

This uniquely human characteristic of living to a ripe old age demonstrates that older people did not lose their value to society even when they were no longer physically robust or fertile. Rather, older people continued to play a vital role in the dense social network that serves as the foundation of culture. For a species that thrives on information, long-lived people are an important resource, particularly in ancient societies, which lacked writing. Older individuals become "living encyclopedias" who can provide useful information that might help a society get through an unusually hard winter or drought, for instance, or some other calamity that might occur only once in a generation. The survival of parents past their childbearing age is also advantageous because they can rear their last children to adulthood, and an important resource in helping to raise grandchildren. While older people typically require more support and care as they age, the advantages of their cultural input must have exceeded any extra costs in food or resources they added to the community at large. "It is clearly worth asking whether knowledge of hunting and gathering techniques, healing practices, ceremonies, social relations and kinship, distant territories—and complex systems of meaning, demanding the use of body decorations—might have been qualities for which these older people were valued," argues White. "This would have been a radical step in the evolution of hominid behavior."

Such a dramatic change in social structure would have had a profound effect on how our modern human ancestors interacted with the environment—and with the other people around them. The Neanderthals may well have lived in a social structure reminiscent of chimpanzees today, with small groups of related females and their young foraging for food in small local ranges while groups of males ranged over larger areas, perhaps hunting larger animals. From about 100,000 to 15,000 years ago, an Ice Age gripped Europe, steeping the terrain in glaciers and a chilly climate. Faced with the increasingly patchy resources in Europe during these periods of heavy glaciation, the Neanderthals may have responded by concentrating their ranges within the few areas in the landscape where resources were diverse enough to live on.

Modern humans, on the other hand, responded to the increasingly dispersed resources in the surrounding environment with a new kind of social organization: They divided up the work among various people, shared food with each other, and had both parents care for the young. The benefits of such cooperative relationships are many, with the result that our ancestors thrived throughout Europe and spread around the world, while the Neanderthals eventually died out.

The effect of this shift to a society that relied on shared resources and division of labor was the creation of an information-based "social economy" where each person existed within an interactive network of rights, responsibilities, customs, and status. Most important for the creative explosion, our ancestors developed a social environment where individuals began to define themselves by their relationship with the groups of people around them, as part of an extended family, clan, or tribe. It is precisely this need to fit into the larger social matrix of society that fueled the explosion of decorative art.

The propensity for modern humans to approach their world in terms of social alliances colored their whole way of life—including how they interacted with their physical environment. Ethnographic studies of modern Eskimo society, for instance, suggest that Eskimo hunters view their prey as participants in a form of social exchange between themselves and nature, and their culture contains many rituals that celebrate the hunt as a social "transaction." Likewise, the magnificent cave paintings of our ancient ancestors may well have been created in an attempt to forge a social relationship with an entity that was abstract—but just as vital—to our ancestors' survival: Nature itself. Creating a thing of beauty as part of a cooperative bargaining with Nature or God is a trait common among human cultures. As one British researcher put it, "Anybody who has lived through an English winter can see the point of building Stonehenge to make the sun come back."

A CLIMATIC TRIGGER

The shift in how our ancestors got along with each other came as the population of *Homo sapiens* grew. At the same time, Earth's expanding glaciers sent a wave of cool temperatures throughout the world, making our ancestors more reliant on each other to survive. Casual relations between various members of a group—and different groups within a region—became more intense, cooperative, and competitive, increasing the social complexity within groups and among different groups in the same area.

An intriguing example of how shifts in climate might have triggered changes in social complexity comes from an ancient site in Africa known as Howieson's Poort. The tools found at Howieson's Poort are so sophisticated in style that most archaeologists would be prompted to regard them as a product of the burst of innovation that accompanied the creative explosion. Yet the tools date back as far as 80,000 years ago, a time when elsewhere in the world anatomically modern humans were still crafting the cruder tools typical of the Neanderthal. More puzzling is that the Howieson's Poort tool style promptly disappeared some 60,000 years ago, only to be replaced by the more primitive tool types of the previous era.

This brief flourishing, then disappearance, of these complex tools may have been caused by a dramatic shift in Earth's climate: About 80,000 years ago there began a period of intense glaciation around the world. Though Africa has no glaciers, the expanding ice to the north nevertheless created a climate change on the continent by shrinking sea levels worldwide, which resulted in much drier conditions in Africa. During this arid period, food would have become scarce, causing environmental stress for the anatomically modern humans who lived there. The sudden appearance of the elaborate tool-making style found at Howieson's Poort came about as our ancestors shifted how they related to each other to survive. As in the time of the Cro-Magnon in Europe some 40,000 years later, the sophisticated, stylized tools found at Africa's Howieson's Poort may have been used as cultural markers that signified a person's affiliation with a particular clan, for instance, and as ceremonial gifts that helped cement social networks, in much the same way that tools are used among modern hunter-gatherers today. When Africa's climate warmed again some 60,000 years ago, the ecological pressure to use these sophisticated tools disappeared. A similar climatic pressure may have led to the explosion of cave painting among our ancient ancestors. Like the Howieson's Poort tools, the height of majestic cave painting coincides with the intense glaciation in Europe that began about 18,000 years ago. After the glaciers began to recede from Europe some 10,000 years ago, instances of magnificent cave paintings disappeared as well.

Thus the psychological seeds for the creative explosion may have in fact appeared with the dawn of the first *Homo sapiens* more than 100,000 years ago. Yet while these people had the capacity for elaborate culture, their creative abilities did not come to full flower until a combination of climatic change and population pressure forced an intensification of social relations, creating an environment where suddenly art, culture, and innovation *mattered*. Before that time, making an elaborate tool or drawing on a rock was merely wasted time and energy. After that time, however, these same activities took on a symbolic meaning to the group-at-large. "I suspect that the origins of art, and body ornamentation, and all this stuff was part of a continual process of experimentation, of play, of generating things that might or might not work," says White. "Someone who lived 60,000 years ago might have occasionally produced a piece of carved bone, but it never went anywhere, because the social structure wasn't there. What dictates whether new things stick is whether there is advantage or application for those new things in the social environment." The groundwork for such a social intensification had already been laid long before within the neural wiring of the brain. The ultimate spark of creativity among our ancestors was *each other*, just as it is today.

A NEANDERTHAL REVOLUTION?

A revealing clue to the difference between having the capacity for culture and displaying that culture comes from a site in France known as Saint Ceasire. At this site, a Neanderthal skeleton was discovered along with a collection of tools that are unlike any artifacts that were made by most Neanderthals during the previous 100,000 years. These Neanderthal tools are very similar to the tools made by the Cro-Magnon, with elaborate blades of flint and tools made with bone and antler—something other Neanderthals did not do. The Neanderthal fossil dates back to only about 36,000 years ago, making it the most "recent" Neanderthal specimen known. The date also demonstrates that Neanderthals were still in Europe thousands of years *after* modern humans arrived on the continent. These sophisticated Neanderthal tools were created some 5,000 to 8,000 years after modern humans appeared.

There is no evidence that the Neanderthals gradually came upon this kind of tool-making technology, and so it is likely that the Neanderthals simply learned, imitated, or picked up the tools from their Cro-Magnon neighbors. Just as modern humans had the mental *capacity* for farming and building computers tens of thousands of years before they actually did those things, the Neanderthals appear to have been capable of producing tools somewhat like those of modern humans—but for as much as 100,000 years they didn't do so.

An analogous leap in technological sophistication recently occurred at Yerkes Primate Center in Georgia, where archaeologist Nick Toth taught a bonobo chimp named Kanzi to make simple stone tools that roughly resembled those of the earliest human ancestor, *Homo habilis*. Though both Kanzi's tools and tool-making ability are a crude imitation of that of humans, Toth's research nevertheless suggests that chimps at least have a rudimentary mental *capacity* to make some kinds of complex stone tools. But while Kanzi's exploits illustrate that there is more to the chimp mind than had been previously thought, it also demonstrates that there is more to tool making than previously thought, too. Tool making involves more than merely having the basic mental and physical ability to chip a rock: Kanzi may have demonstrated that bonobos have the capacity to make stone tools, but, in fact, bonobos do not do so in the wild. Similarly, it may well have been that while the Neanderthals were capable of producing complex tools, they lacked the complex social structure that would make such tools as useful to them as similar tools were to modern humans. It was the invention of the Cro-Magnon's new social structure, not the tools themselves, that made the technology possible.

The idea of having a mental capacity for a kind of behavior without actually doing it would not appear so puzzling, if people didn't make the mistake of equating evolution with *progress.* No single adaptation is clearly "better" than another in the grand scheme of life on Earth: Every change in behavior is a balance between costs and benefits, between long-term and short-term gains and losses. We tend to judge the Neanderthals as failures because they went extinct, but, in fact, the Neanderthals could be thought of as one of the great success stories in evolution. They roamed the earth for more than 200,000 years—a tenure that is equal to or greater than that of modern humans so far. To modern-day eyes the human tool kit may appear to be a clear improvement over that of the Neanderthals. But to the Neanderthals—who had lived successfully for thousands of generations using their "old" technology—changing how they went about their daily lives would have been psychologically plausible only as an act of desperation.

Though the cultural explosion appears to be a huge success from the vantage point of 40,000 years later, it is important to remember that at the time, neither the Neanderthals nor our modern human ancestors had any idea what the long-term results of their changes in behavior ultimately would be. "You can't look at the creative explosion as a set of solutions to a set of perceived problems," says White. "Rather, it was a change in behavior that had enormous consequences that were totally unforeseen by our ancestors. What are the consequences of the beginning of art? Probably the space age. Would our ancestors have seen it that way? Not a chance. What will be the consequences of agriculture? Perhaps the extermination of the human species. Did the people who began to farm see it that way? Not a chance. These things were perceived as short-term solutions to immediate problems, and they end up having consequences that could never be foreseen by the people doing them."

BUILDING BETTER WIDGETS

One early sign of the explosion of creativity in our ancestors about 40,000 years ago is the sudden leap in variation in their tool making. Before that time, the kinds of tools hominids made around the world were fairly simple stone and wood implements of limited design that were unchanging in the face of shifts in the environment. "With the Neanderthals, we see the same type of tool being applied over and over again, regardless of whether it's cold or warm or forested or whatever," says White. "If you look at Neanderthal tools in France and Neanderthal tools thousands of miles away in Asia, they are all the same. But once you hit 40,000 years ago and the arrival of modern humans, you can't even use the same classification scheme to compare what you find in France to what you find in Italy. They are that different."

To make their stone tools, the Cro-Magnon shaped a chunk of flint into a large cylinder about the size of a coffee can, then struck a series of long, thin, slivers of stone off the outer edges of this core. These blades were then used as the starting point for a host of different tools: Chipping away at one end of the blade produced a spear point, for instance, while chipping along the side of the blade produced a knife edge or scraper for animal hides. Using this stone-cutting method, the Cro-Magnon could get roughly forty feet of "cutting edge" from a pound of flint—as opposed to forty *inches* per pound using the older, Neanderthal-style method of stone working. The Cro-Magnon also made stone awls and picks for making clothes out of hide, and pioneered another innovation of everyday life: using bone, ivory, and reindeer antler as a new medium to express their creativity. The Cro-Magnon shaped the tough, plasticlike material that makes up bones and antlers into jewelry, harpoons, fishhooks, and sewing needles. "Antler is a significant raw material that was never used before the arrival of modern humans," says White as he strolls through his lab. "In this single drawer here, I'll show you more spear points made of reindeer antler than existed in the *whole world* prior to the cultural explosion."

A former student of White's, Heidi Knecht, discovered that the Cro-Magnon used an ingenious method of fastening spear points made of antler onto wooden shafts: "It's something that you or I would have never thought of," says White, pulling out a tray that is filled with dozens of ancient spear points. "Compared to the Neanderthal tools, it's the space age. You take an antler, split it in half, and you wear it down into a point. Then you put a split in the nonpointed end. Ever since archaeologists began finding these spear points they thought, 'Well, that's easy: The Cro-Magnon stuck the split end onto a flange at the end of a wooden shaft.' Well, it now turns out that this is not at all what they did. They took a wooden spear shaft, split it, and then put the split end of the point into that groove. Then they drove a tiny wedge between the flanges to spread them apart within the groove, so the point was stuck firmly. Obviously, this is not the simplest way of doing things. I mean, *this is us*."

MAN THE HUNTER-FOR-REAL

With the modern human's new tools came a leap in sophistication in hunting, revealing that for the first time, our ancestors had mastered the art of working together to bring down huge amounts of meat on the hoof. The idea of early humans as big game hunters has a venerable legacy in the study of the evolution of the human psyche. For many years the notion of "man the hunter" was the prime candidate as the "trigger" of human evolution. After all, hunting down large, ferocious beasts required sophisticated tools and a big brain to wield them with. But, in fact, the image of our earliest ancestors being big game hunters is largely a myth.

To be sure, our modern-day bodies reflect the evolutionary legacies of a long history of eating meat. It is not only the big human brain that distinguishes humans from our evolutionary cousins, the apes; our bowels do, too: The gut of an ape is dominated by the large colon—the big, tubelike organ that helps process the tough fibers of plant food—and this testifies to the ape's everyday diet of massive amounts of vegetation.

The human gut, on the other hand, is unique among primates in that it is dominated by the small intestine, which is where protein and nutrients are rapidly broken down and absorbed—an anatomical arrangement that suggests that humans have long been eating nutrient-rich foods such as fruits, nuts, and meat. Further evidence that this dietary preference stretches far back into antiquity comes from the fossil teeth of our ancient human ancestors, which have only a thin coating of enamel and show none of the grit marks and heavy wear that would result from a diet consisting solely of tough plant fibers. Also, the human body can't make vitamins A or B12, two vital substances that are commonly found in meat.

But other quirks of our modern-day "eating anatomy" reveal that the meat our ancestors typically ate was by no means like the sirloin we consume today. The human body absorbs 95 percent of the fat we ingest, suggesting that this compact source of calories was extremely hard to come by in ancient times. This legacy of our ancient ancestors' eating has created a psychology of food preference that is celebrated in the local burger palace. Having existed for eons in an environment where only lean meat was available, our ancestors would have found a fast-food hamburger a gustatorial paradise. It is precisely what our ancestors loved about fat—its incredibly rich content of calories—that makes it so bad in modern times, when fat is available in great quantities. This evolutionary legacy in our food psychology is the reason that fatty foods cause so much trouble for those of us who live in the food-rich industrialized West today: Having evolved in an environment where fat was scarce, our modern-day minds have a hard time knowing when to stop.

Meat is the ideal food for the most cooperative species on Earth: Packed with calories, meat is one of the most compact sources of nutrition going. In every type of environment from savanna to Arctic, hunting large game provides the most calories per hour of work among all the various ways of obtaining food, regardless of whether people are using stone-tipped spears, nets, fishhooks, or bows and arrows. Thus meat is a great bargaining chip for social relations. A person could typically gather and carry about enough plant food for at most only a few people to eat. The carcass of a large mammal, however, contains enough food for many people, making it ideal for sharing with friends or neighbors, courting lovers, tending to the sick, or provisioning a family.

Incorporating more meat into their diet no doubt created a whole new set of problems for our ancient ancestors. For one thing, unlike plants, which are firmly rooted in the ground, meat has the propensity to run away, making it a less reliable source of food and requiring more sophisticated methods of hunting, trapping, or otherwise getting it. Hunting or scavenging meat for a living also brought our ancestors into direct competition with some very nasty creatures such as big cats and hyenas.

THE BIG SWITCH

The first sign that our ancient ancestors had begun to make the changeover to including meat in their diet comes with excavations of the archaeological sites of the 2-million-year-old ancestor *Homo habilis.* One of the most famous places where this ancient ancestor plied its trade is Olduvai Gorge in Tanzania, which is often called "the Grand Canyon of prehistory." At one site there, dating back about 1.85 million years, archaeologists uncovered several huge piles containing more than 2,500 stone tools and some 15,000 pieces of animal bones ranging from mice to pigs to antelope to elephants.

In keeping with the old "man the hunter" image of our forebears, anthropologists at first assumed these piles of stones and bones were the remains of campsites left behind by roving bands of hunters. Thought to be evidence of "home bases" not unlike those of modern hunter-gatherers, the sites suggested that early humans had a life-style where men hunted for meat as women gathered plant foods. Later these ancient ancestors would meet back at their home base and share their food with others, perhaps engaging in the beginnings of the social and cultural practices that characterize modern human culture. A circular arrangement of stones was uncovered at Olduvai that resembled the remains of the twig huts built by modern hunter-gatherers, and so was proposed as evidence of early home building.

As attractive as this vision of our ancient ancestors' life was, however, archaeologists discovered evidence of a very different life-style when they began to look more closely at the stones and bones *Homo habilis* left behind. Putting the bones under an electron microscope, researchers found that many of the ancient bones bore unmistakable signs of cuts and gouges made by stone tools, in much the same way that a wooden kitchen cutting board is scored by the cutting action of a knife. The cut marks are clear evidence that our ancient ancestors were using tools to cut meat off the bones in the sites at Olduvai.

But while the powerful vision of the microscope demonstrated that our ancestors were eating meat nearly 2 million years ago, it also revealed other marks on the bones that made the notion of our ancestors living like miniature Boy Scouts at home bases a lot less likely. Many of the bones bore scratch marks, for instance, that were made by the teeth of carnivores such as hyenas, suggesting that animals other than humans *also* ate the bones at the Olduvai sites. Other scratch marks on the bones revealed that, at least part of the time, our *Homo habilis* ancestors were not hunting but *scavenging* for their meat. On some bones, scratch marks made by stone tools cross over marks that had previously been made by carnivore teeth, and on other bones there are teeth marks that cross over tool marks. Thus both carnivores and hominids appear to have eaten the same bone. On about half these doubly cut bones the hominid made the cut first, suggesting that the carnivore got to the bones after the hominid discarded them. On the other half, however, the carnivore tooth marks appear first, suggesting that our ancient ancestors had scavenged the bone from a carnivore kill site.

SOCIAL SUPPERS

Real big game hunting didn't appear until the cultural explosion millions of years later. Archaeological evidence reveals that modern humans became quite adept at bringing in all manner of beast and fowl. The early modern humans' diet included big game such as deer, bison, oxen, horse, boar, and mammoths. They also ate a variety of birds such as ducks, pigeons, and grouse, and they trapped foxes, beavers, and rabbits for their pelts. Indeed, the reason that many species of large mammals no longer walk the earth today (other than cows and other domesticated animals) is that our ancient ancestors *ate* a good many of them. "Mammoths went extinct all across Europe by 12,000 years ago," says White. "It wasn't their fault and it wasn't the environment's fault. It was humans."

Our modern human ancestors' hunting abilities are strikingly apparent in the Americas, where Paleoindians hunted to extinction nearly *70 percent* of the species of large mammals on the continent, including mammoths, camels, and giant sloths. The history books may say that horses were introduced to North America by the Spanish during the 1500s, but horses were in fact native to the continent—and had been hunted to extinction some 10,000 years before the Spaniards' arrival.

The most important factor in our ancestors' hunting prowess, however, was not their new tools, but their psyche. Large, relatively clumsy, and slow of foot, our ancestors could not stalk their prey and swiftly chase them down, like many predators, but instead had to rely on working together as a team to bring down their prey—something at which modern humans excelled. They used fire or noise to flush their quarry out of the bush while other hunters lay in ambush for the fleeing animals. Opportunities for ambush also came as our ancestors made their camps deliberately in the paths of herds of caribou as the animals followed their annual migration routes. More than half of the early human archaeological sites in southern France, for instance, were situated less than a mile from a river or stream. Most of the sites appear at places where the water in the river is low, making them an ideal crossing point for reindeer.

Further evidence of our ancestors' extensive knowledge of the lives of their prey comes from a site in eastern France known as Le Roc Solutre. The area contains the remains of more than 100,000 horses, many of which show signs of having been cooked. When the site was first discovered more than a century ago, it was assumed that early humans stampeded the animals off a nearby cliff, much as Native Americans ran bison off high bluffs in mass slaughter. But the horse remains in fact do not lie directly beneath the cliffs—they are, however, in a cul-de-sac off a wide valley that cuts through the mountainous ridges in the region. Knowing the horses would be passing through the valley as they migrated from their summer range toward their winter range, early humans apparently set up a barricade that herded the horses into the cul-de-sac, where the hunters could easily ambush and spear their prey.

The biggest leap in the everyday lives of our ancient ancestors that came with hunting was not so much in how they got their food, but in the kinds of cooperative, social exchanges they made once they got it. At one site, archaeologists determined that a single carcass of the same animal had been divided among three different fire hearths that were hundreds of feet apart. The find suggests that the beast had been shared by three different families—a practice echoed in our lives today, from a potluck dinner to a Thanksgiving feast.

ART: THE SOCIAL SIGN

The dramatic changes in social structure that accompanied the creative explosion is reflected in our ancient ancestors' crafting of vast amounts of "personal" art. As compelling as the famous ancient cave paintings left by our ancestors may be, they in fact represent only the *midpoint* of "art history" of the human species to date. Our ancestors' first artistic efforts were created nearly 15,000 years earlier, and they were not executed on cave walls but were carried and worn as decoration, as beads, pendants, and statuettes.

Using samples of elephant ivory that government officials had confiscated from smugglers, White has recreated the process by which ivory beads were made by modern humans some 35,000 years ago. "Using modern materials to replicate ancient human behavior is the cutting edge in archaeology," he says. "We want to know how people 35,000 years ago could produce these finely polished beads, given the materials and technology they had available to them. So we take the materials and technology they had back then and try to do it ourselves. And then we go back to the original objects and look through the scanning electron microscope and compare what we see on the original beads with what we produce. And guess what we see? We've replicated it."

White found that producing these ivory beads was no mere idle pastime, but a time-consuming, arduous task. First, the ivory tusk of a mammoth would be split lengthwise into small rodlike "blanks" about the size and shape of an unsharpened pencil. Circular grooves were then scored around the circumference of the rods, cutting deeper and deeper until small, aspirin-sized beads could be broken off one by one. A hole would be gouged out of a bead with a sharp rock, then the bead would be scraped and filed into a teardrop shape. The beads were then either hung on a string or, as the patterns of beads found in ancient gravesites suggest, sewn onto clothing. By gouging the hole first, then wearing down the material around it, Cro-Magnon artisans could produce the beads' delicate, uniform shape without the aid of a drill, which apparently was not part of the early Cro-Magnon tool kit.

Contrary to the popular notion that art is a mere by-product of a society where people's vital needs have been satisfied—and so people have plenty of leisure time to sit around and be creative—White's reconstruction of the bead-making technology suggests that art was as crucial a part of our ancient ancestors' survival as finding food and shelter. "These beads were very important to these people," he says. "They spent thousands of hours of their life making ivory beads even though they were living within 150 kilometers of a huge glacial ice sheet." At one 28,000-year-old burial site, researchers found a total of *14,000* ivory beads, each one of which would have taken some forty to fifty minutes to make. For our ancestors to have committed so much time to making art while the world was locked in the extremely cold, harsh conditions of an Ice Age climate testifies that art was not some simple pastime, but rather must have played a vital, important role in early human society. "People don't often think about art as a necessity," says White. "But it seems to have arisen as an attempt to cope with some very difficult circumstances. Art corresponds to periods of stress much more than it corresponds to periods of well-being and leisure. There is no question that art is not being produced as a result of people having more spare time; it is being produced when all hell is breaking loose and people don't have enough to eat."

Art was important for dealing with such a difficult climate because it plays a role in people's cooperative interactions with each other—interactions that were, and still are, vital for human survival. When people adorn themselves with various art objects, their bodies become a "living canvas" on which various signs, signals, and expressions of personal views and associations are displayed. This is certainly true in modern society, where a wedding band, a fan's baseball cap, or a gang member's "colors" express one's affiliations and loyalties. The "artwork" that people wear becomes a visible display of their character, inviting others to interpret their meaning. Our modern minds are quite prone—sometimes dangerously so—to leaping to conclusions about someone's political views, financial status, upbringing, social strata, trustworthiness, and overall character from body decorations. From tattoos to Phi Beta Kappa keys, from spiked heels to huge diamond rings, the variations in the kinds of adornment men and women display provide clues to their views and status in a society. Likewise, the various portable and wearable art objects made by early modern humans no doubt served as part of the social glue that helped weave their society together. For instance, in some regions in Europe, archaeologists find ancient tools that are decorated with nearly identical animal carvings, suggesting that this image may have served as a totem sign that identified the toolmaker or the person carrying the object with a particular clan.

Just as in modern society, where a Rolex watch tells far more than the time, our ancestors' personal art no doubt also advertised a person's prowess and achievements—thereby making known one's worthiness as a cooperative partner or threat as a powerful adversary. Nearly one-third of the pieces of ancient jewelry archaeologists have found, for instance, are made from carnivore teeth from animals such as foxes, wolves, and lions. At one burial site, for instance, there are more than 240 fox canines around one person's waist—the equivalent of several dozen foxes. Yet carnivore bones make up less than 10 percent of all the animal remains typically found at early human sites. "It is surely more than just a coincidence," says White, "that animals that hunt other animals were singled out for use in social display by the most dangerous predator of all."

Another function of portable art may have been to build alliances not only among a single group of people but also among the various bands of people who populated Ice Age Europe. Early modern humans had extensive networks and alliances through which goods were traded and distributed throughout the continent. Nodules of high-quality flint, exotic seashells, mammoth ivory, and amber have all been discovered at ancient sites that are hundreds of miles from the materials' sources. Artifact-rich sites such as Spain's Altamira cave may have served as a kind of "convention center" for large congregations of different groups of people coming from miles around. The great diversity of styles in the personal ornaments found at the cave suggests that at these gatherings personal art objects were exchanged between groups to cement cooperative alliances or make peace, much as modern societies exchange gifts for social purposes today. One enigmatic feature in the personal art of our ancient ancestors are the numerous "Venus" figurines that have been found throughout Europe from France to the Russian plain. Sculpted from stone or etched in relief, and dating back tens of thousands of years, these Venuses share an archetypical design of exaggerated female features: wide hips, tapering legs and shoulders, large breasts, and nondescript faces. The Venus sculptures appear to be the first widespread artistic tradition among early humans, perhaps serving as an icon for a religious or fertility cult.

SUBTERRANEAN CATHEDRALS

The most magnificent examples of ancient creativity are the majestic paintings our ancestors made on the walls of caves, the most famous example of which is the caverns of Lascaux in southwest France. Discovered by four boys a half century ago, Lascaux contains hundreds of paintings of animals: Wild horses gallop down the walls of one chamber past a giant auroch—an extinct cattlelike creature—to confront several stags and more aurochs head-on. Farther into the cave are images of bison, deer, horses, and a half-dozen lions, all exquisitely painted and in such naturalistic positions that they must have appeared alive when viewed by the faint flicker of torchlight. Deep within the cave is one of the most enigmatic figures of Ice Age art: At the bottom of a twenty-six-foot-deep shaft of rock is a painting of a huge bison, its head and horns pointed downward, poised to attack a curiously drawn, almost sticklike figure of a man wearing what appears to be a bird mask. The bison has been wounded by a spear, its entrails spilling, and the man has fallen in front of it. Nearby is what appears to be a tool adorned with another bird image.

The haunting image of fallen hunter is indicative of the deep mystery of Ice Age art as a whole. One can't help looking at the images and struggling for understanding: What do the paintings mean? What were the artists trying to convey? What role did the art play in society? Not surprisingly, perhaps, these questions are still being asked about art being created today.

One suggestion is that the cave art was part of some kind of ritual in which Ice Age hunters used the painted images in ceremonies to increase their chance at a hunt. A statistical analysis of nearly 2,000 images of animals found in another cave in France suggests that the aggressive behavior of certain animals may have played a role in whether they were depicted on cave walls. Dangerous animals, such as horses, bison, and ibex, were pictured in the cave art more frequently than their skeletal remains appear in nearby camp sites left by the early humans. This suggests that the Ice Age paintings don't merely reflect our ancestors' success in the hunt, but that the animals' reputation as fierce prey may have had a symbolic meaning that influenced their being portrayed in a painting.

Further insights into the role of cave art in early human society has come from new efforts to analyze the pigment used in the paintings. In one recent study of some seventy-five paintings from the famed Niaux cave in France, researchers found that the paints were made with combinations of ocher, charcoal, and other substances that were combined in characteristic "recipes." Often the paint itself can be dated by conventional archaeological methods, and, in one instance, a recipe of paint used in several paintings matched that of a daub of paint that was found on bones discovered at a nearby ancient campsite, which could be dated. The results reveal that paintings at Niaux, which researchers had long assumed had been painted at roughly the same time because they were painted in a similar style, were actually painted several *centuries* apart.

The endurance of these images across generations suggests they were not just pretty pictures, but symbolic icons that persisted in our ancient ancestors' culture for centuries—much as certain icons persist in society today. "Think of the crucifix and what it stands for," says White. "When you put it on a wall in a twentieth century building, you are evoking events and philosophies that were developed 2,000 years ago." Pigment analysis reveals that while most of the animal figures in the Niaux cave appear to have been painted more or less spontaneously, the animals in the cave's famed "Salon Noir" were first sketched out in charcoal, then painted over. The careful, deliberate planning of the artwork suggests that this part of the cave was a type of "sanctuary," a special place where the artwork was carefully planned, perhaps for ceremonial purposes.

There are other hints that these cave paintings may have played a role in rituals performed by shamans or spiritual leaders: Found among the realistic images of bison and deer on the caves' walls are paintings of strange half-animal, half-human figures, as well as enigmatic spots, chevrons, and other cryptic markings. These wall markings may depict the images shamans see when they enter a hallucinatory trance: Neurological research indicates that trancelike states can often cause people to see visions of dots, lines, and curves—in the same way that a person sometimes sees "stars" after a bump on the head. This connection between art and spiritualism is a part of human culture that continues today: Many of the paintings of modern- day hunter-gatherers known as the Kalahari San, for instance, were generated by shamans who enter powerful trances to try to cure the ill or communicate with the spirit world.

Perhaps the most famous example of cave art that suggests some sort of shamanistic, religious practice is the twin caves of Trois Frères and Tuc d'Audoubert in France. The caves lie within fifteen feet of each other and were once connected underground. The main chamber of Trois Frères lies deep within the cave and is reached by a tortuous route that is dotted with animal drawings. In the main chamber there are many engraved animals but only one painting: Perched high on a wall and dominating the chamber is a figure that is known as the "sorcerer"—a half-man, half-animal figure with owl-like eyes and stag horns sprouting from his head. Unlike most of the paintings of Ice Age art, the sorcerer's gaze turns outward, its eyes fixed with a trancelike stare. Some researchers interpret the figure to be a shaman dressed in ritual costume. A similar half-human, half-bison figure is also engraved nearby, and in a hidden passageway there is another "sorcerer" figure who appears to be playing some sort of musical instrument.

At Tuc d'Audoubert is an equally enigmatic setting that suggests some sort of intense ritual among the ancient people who gathered there. Af-

ter navigating a tortuous passageway through the cave, a visitor is confronted with a series of strange artifacts on the way to the main chamber: A stack of limestone plaques, many with figures drawn on them,
appears first, followed by the carefully arranged remains of several bears,
next followed by the remains of a snake. Then, on a ledge near the wall,
lie the tooth of a fox, a bison, and an ox, each pierced and decorated with
ocher. In a small side chamber no more than five feet high are some fifty
tiny footprints—heelmarks, actually—in a distinctive pattern that suggests that children were performing some sort of ritual dance or proces-
sion. In the main chamber itself there is no art at all, save for a
magnificent clay sculpture in the center: Two bison, each about two feet
wide, lie in remarkably lifelike detail.

Imagining the conditions under which these artworks were viewed
some 15,000 years ago suggests the potentially awe-inspiring nature of
the experience. Unlike modern-day archaeologists, our ancestors viewed
these figures by the dim flicker of fat-burning lamps, having reached
these remote corners of the cave via a difficult if not dangerous journey.
Caves by their very nature suggest an other-worldly existence, consisting of often claustrophobic, disorienting labyrinths that take a visitor farther and farther away from sunlight, vegetation, and other trappings of
earthly experience, and whose shimmering walls and exotic rock formations naturally suggest creatures frolicking—or lurking—in the shad-
ows. Yet at the same time, caves can be warm and even womblike
sanctuaries that protect against the harsh, Ice Age climate.

The role of Ice Age art in helping to weave together the social cur-
rents of ancient society should come as no surprise, for art continues to
play that role in society today. From graffiti to television to film to painting and sculpture, art continues to shape our vision of who we are as individuals and as a society. Likewise, the Ice Age paintings served as
"social glue"—shared images that bound together our ancestors through
common experience and iconography that spanned generations. "Art
transcends human lives," says White. "Whether it's painted caves, or engraved stones, or little statuettes; these things may have circulated for
hundreds of years as manifestations of ideas. It's what anthropologists
call 'social reproduction'—the tacking on of knowledge, ideas, and be-

liefs from one generation to the next. It couldn't have been done without these objects that transcended generations: They are like people: They have histories; people talk about them."

Our ancestors' creation of artwork signaled a leap in their social and cultural complexity that had down-to-earth, practical applications. "Art is a cultural equivalent of mutation," says White. "We know that bipedalism was caused by a genetic mutation that allowed some creatures to be able to stand upright, and that proved advantageous. What about the individual or social group that creates drawings of animals? What is the adaptive value of that? Well, for one thing, it creates the ability to communicate with each other through images: People recognize that they can represent ideas, creatures, objects, what-have-you, with forms. Once you can do that, you can talk about them without their being there. You can conceptualize things. You can create a tool that doesn't exist by making a drawing of two tools that do exist. Where would engineering be today without drawings? We've tended to think of the imagery from this period as 'art' in the soft sense that people were into evocative images, and animals really meant a lot to them, etc.—and maybe they did. But you also have people who are capable, using artwork, of showing someone else how an animal behaves, or where best to plant a spear, or any number of practical applications."

ANCIENT MELODIES

If Ice Age cave art indeed played a role in our ancient ancestors' social rituals, these ceremonies were no doubt accompanied by another quintessential aspect of human psyche: *music*. Music appears to have been an important part of our early ancestors' psychology. At some archaeological sites, researchers have found tiny pieces of bone and ivory, decorated and pierced with a tiny hole at one end. Tied to a string and swung in a circle, they make a loud hum akin to a modern-day bullroarer. The fragmented remains of whistles and flutes made from the bones of birds have also been found, as have been a set of painted mammoth bones that may be drums or an Ice Age xylophone.

The Ice Age paintings that grace the caverns may have been purposely placed to take advantage of the remarkable acoustic properties of cave walls, which can echo or dampen noise depending on where a listener stands. Whistling through five octaves as they walked slowly through three caves in the Pyrenees, researchers mapped the places in the caves where the rock formations produced the best acoustics. Comparing their acoustic map with a map showing the locations of various artworks, the scientists found that all the prime locations for listening to sound were accompanied by some kind of artwork, even if the nearby cave wall surface was so small it had only room for a set of painted symbols, for instance. In contrast, those places with poor acoustics had very little art nearby. Thus, the best places to view the artwork of the cave appear to have also been the best places to hear music or chants.

That our ancient ancestors' creative culture would include music is not surprising, for the ability to create and enjoy music is a deeply ingrained part of the human psyche. All human cultures have some form of music, and some features of how we perceive and organize musical tones are a universal trait of the human species: The ear is capable of discerning hundreds of minuscule differences in the pitch of a sound, for example, yet nearly all cultures typically divide this vast range of audible sounds into musical scales of only about five to seven notes. Tellingly, a replica of a bone flute that was discovered at one ancient cave site produces a similar five-note scale. Further evidence that the brain is specially wired to enjoy music comes from people who suffer brain damage from stroke and are afflicted with "amusia"—an inability to recognize familiar melodies and loss of musical ability—even though other mental abilities are left unimpaired. The wiring up of the brain's musical knowledge begins very early in life and, like language, is "tuned" to a particular culture. Six-month-old infants possess a rudimentary ability to perceive that a musical chord contains a "sour" note that is atonal. By age one, North American children are better at remembering a melody when the tune is created from notes in a scale found in conventional Western music, as opposed to melodies written from a more exotic scale used in Indonesia.

Why should an appreciation of music be part of the burst of new behaviors that blossomed during our ancient ancestors' creative explosion? Unlike painting or sculpture, music typically doesn't mimic anything found in everyday life. There are exceptions, of course: Richard Wagner incorporated the sound of hoofbeats into his *Ride of the Valkyries,* Bach recreated the sound of an earthquake in his *Saint Matthew Passion,* and some other orchestral works include cowbells, sirens, and other sounds from the everyday real world. But mostly, music sounds like itself, and as Plato observed, "It is very difficult to see that any worthy object is imitated by it."

New research suggests that while music doesn't directly copy any sound in nature, in more subtle ways music reflects the workings of the human mind. Studies of the physics of sound suggest that the brain may enjoy music because the overall organization of a musical piece—be it modern, classical, Western, or foreign—in a fundamental way echoes the structures found in natural formations such as mountain ranges, trees, and perhaps even the brain itself. Nearly all kinds of music around the world, old and new, share a fundamental mathematical formulation that describes how the musical tones in a piece relate to each other. Playing a series of notes that are randomly selected from a piano keyboard, for instance, results in a sound that is best described as noise. Likewise, merely randomly sliding up and down the piano keys—such as when you are dusting off the keyboard with a rag, for instance—produces a sound that is so predictable and monotonous that it, too, would not qualify as music. What our species calls *music*—for this applies to nearly all music from all human cultures—consists of notes whose order lies halfway between the randomness of noise and the dull predictability of a simple up-and-down movement of notes.

This same mathematical mixture of randomness and predictability is present in the patterns of nature, too: The ebb and flow of river banks, variations in the beating of the human heart, the electrical activity of the brain's cells, the branching shape of a tree, lung, or river delta, all share this mixture of random and predictable variation. In fact, this same kind of mathematical formulation—known as a *fractal*—is used by computer scientists to generate amazingly realistic computer images of coastlines, clouds, mountain ranges, and other natural scenery. The consistency of this mathematical formulation throughout nature is responsible for the fact that a mountain range viewed from an airplane looks just like a wrin-

kled blanket viewed from above one's bed: Far from being a random as-
semblage of features, Nature is woven together in a way that has an order and rhythm that exist throughout the world and at every scale—an
order that is reflected in the way the pattern of notes in music changes,
too. Indeed, several modern composers are experimenting with creating *fractal music*, using the equations of nature to generate music that
sounds remarkably like the melodies that come from the human mind.

It may be that the ancient Greeks, who believed that music was a reflection of the "harmony" of the planets, almost got it right. People enjoy music not because it mimics birds, waterfalls, or other natural sounds
but because the mind's musical abilities evolved to be in tune with the
ultimate patterns of nature. The joy of music lies in the fact that it reflects the rhythms of life itself, and an appreciation for those rhythms is
an ingrained part of every person's psychology.

Like art, music no doubt helped to knit together our ancient ancestors' society, just as it continues to do so today. Not only does music form
a common dialogue of shared culture, but research also has shown that
most people have a subconscious sense of pitch that they use in interactions with each other in everyday life. This uncanny ability to place
the pitch of a sound within a range of sounds that are common in a community serves as a kind of warning bell or "lie-detector": People who
are under stress often unconsciously speak in a voice that is pitched
higher than normal. The similarity of sounds both heard and produced
by the individuals in a community also serves as subtle cues that help
bind people to the group as a whole and, like a person's accent, helps signal that a person is part of one's common culture or a stranger from somewhere else. In other words, our perception of music is part of a larger
strategy we use to create and maintain social relations through the shared
experience of sound.

THE MODERN EXPERIMENT

With their explosion of culture, our ancient ancestors created a new
social structure that drastically changed the way they went about their
lives. In doing so, they set the human species on a trajectory that was to
go through another dramatic social upheaval tens of thousands of years

later. The evolved mental mechanisms that we use for art, language, choosing mates, coping with violence, cheater detection, and a host of other social skills were forged while people lived as small groups of hunter-gatherers. But about 10,000 years ago, our species began a sudden shift to living in large, sedentary groups that sustained themselves through agriculture. In doing so, they created a whole new social environment for people—and a host of new problems that continue to challenge us today. Can a psyche that evolved in the Stone Age cope with the modern-day world?

Chapter Nine

THE MODERN

EXPERIMENT

Civilization is looped together, brought under a rule, under the semblance of peace by manifold illusion.

—WILLIAM BUTLER YEATS

W hen the Old World sailed across the Atlantic Ocean and encountered the New World nearly five centuries ago, each no doubt experienced an eerie shock of recognition. The Spanish Conquistador Francisco Pizarro and his tiny band of soldiers expected to find savages in a savage land as they made their way through the highlands of what is now Peru in 1532. Instead, they found themselves in the midst of a remarkably sophisticated, complex society that in many respects was not very different from the one they had left behind in Europe.

Even though the Inca lacked writing, the wheel, and iron tools, their civilization still must have seemed strangely familiar to Pizarro and his men. Cradled between South America's Pacific coast and the towering Andes mountains was an empire ruled by a powerful monarch and filled with aristocrats, government workers, artists, priests, warriors, and peasants. The Andean hillsides were lined with terraces of fields watered by an elaborate network of irrigation canals. A system of paved roads spanning 14,000 miles—with some 10,000 state-owned "rest stops" stocked with freeze-dried food scattered along the way—linked a population of 12 million people who lived in an empire that stretched farther than that of the ancient Romans. Cuzco, the Inca capital city, was "so beautiful and has such fine buildings," wrote one chronicler to the Spanish monarch, King Charles, "it would be remarkable even in Spain."

The Andes region is one of six places in the world where, starting about 6,000 years ago, humans suddenly began settling down and living in large, complex societies. The other "cradles" of civilization include Mesopotamia—where the first large, complex societies appeared—and the Indus Valley, Egypt, China, and Mesoamerica, where the Maya and Aztec societies developed. Though there was no doubt contact among some of these societies, these six centers of civilization developed independently. That is, all over the globe and at roughly the same time, people began to settle down into large, sedentary groups.

The shift to living in large-scale societies created enormous challenges for our Stone Age mind. For hundreds of thousands of years, our psychology had evolved in a familiar setting: Social relations among our ancient ancestors were carried out primarily within small groups consisting of extended families and close, lifelong associates who roamed the land, taking their scant belongings with them as they followed herds of animals and collected plant food. Settling down into large groups meant a host of new problems for a mind that had evolved to cooperate with a small number of familiar friends and relatives. Suddenly, people were forced to rely on strangers for their vital needs—often without the family ties or repeated exchanges that help cement such relationships. Because food was farmed on a particular piece of land and stored—not just gathered day to day—a person's livelihood became vulnerable to being taken by others, creating greater need for mutual defense and breeding fear and suspicion of strangers. Being dependent on one plot of land for survival meant that people could not simply move elsewhere if a dispute arose between spouses or neighbors, as had been the practice for most of human prehistory.

Viewed from the perspective of the evolution of the social mind, the rise of civilization was not quite the great leap forward of progress toward the pinnacle of human existence that has long been portrayed. Certainly, the development of modern medicine and science, the invention of writing, the establishment of laws, and the creation of elaborate artwork and sophisticated architecture are all impressive and have improved the lives of many people in the modern world. But being part of a complex society is not inevitably all to the good, nor is civilization something that, once invented, was unhesitatingly adopted by people everywhere who recognized its many virtues. Rather, living in a large society is a delicate balance of costs and benefits that sometimes results in some people's lives being worse, not better. The rise of civilization represents not a march of inevitable progress but a desperate shift in human behavior that was an attempt to survive amidst challenging new surroundings. The changes brought about by the social upheaval caused myriad problems, and continue to pose the ultimate question of whether the human species can survive.

IS CIVILIZATION GOOD FOR US?

Living in huge, complex societies poses an enormous challenge to our Stone Age mind. It is difficult to play TIT FOR TAT with a great number of people, for instance, because it becomes harder to mete out punishment for a defection. Unlike the small groups that our ancient ancestors lived in for most of human existence, the large groups of modern civilization make it less likely that someone who "defects" on you will ever meet up with you again. In cases such as a random mugging, a corporation who dumps toxic waste in your neighborhood, or the proverbial used-car salesman, you may have never met the defectors previously, nor will you see them again. Plus, in many cases, the costs of tracking down the defector and retaliating—and the potential for further retaliation by the defector—make punishing a defection a more difficult task. Hence the classic urban rationalization for ignoring the street criminal who brazenly smashes a car window and yanks out a stereo in plain sight of dozens of people: "I didn't want to get involved."

To study how the cooperative behaviors that served our ancestors so well fare in the strange, new social environment of a large-scale, anonymous society, University of Michigan's Robert Axelrod conducted a series of computer simulations that were similar to those he did in his study of the evolution of cooperation. His research on how cooperation evolves among large groups of people is an example of an emerging trend toward using computers, mathematics, and game theory to study the evolution of culture. In the same way that a biologist might study how the physical traits of animals evolve, researchers are examining culture as a collection of thoughts or icons that undergo a kind of evolution within a society. These cultural icons are sometimes called *memes* or *cultigens*, and they evolve within the collective minds of people in a society in a manner a little like their biological counterparts, genes: They compete with each other for survival, multiplying or dying out as their value to the group changes. These memes might be a new type of tool, a new technology, a new joke, or a new concept such as the Bill of Rights. Unlike genes, these cultural adaptations do not depend on biological reproduction to multiply, and so can quickly spread throughout a society, and can be easily varied through experimentation and innovation.

This new brand of research is being applied to the question of how "standards" of behavioral norms evolve in a society. These societal standards dictate what is or isn't proper behavior in a group, from "Don't yell in public places" to "Don't murder," and therefore help people in society overcome the challenges of cooperating in large groups. One problem with maintaining cooperative behavior among a great number of people is that often the costs to society of a single defection—cutting into a long line, cheating on one's income tax, or dumping one's trash into a lake—are spread out among a large number of people. No one is hurt very much by any single defection, even though over time the group as a whole suffers.

This predicament is known as the "tragedy of the commons," after a famous example that captures the essence of the dilemma of group cooperation. The "tragedy of the commons" concerns a group of sheepherders who graze their flocks on a communal pasture or "commons," which is shared by all. One day a shepherd decides to add a sheep to his flock. The impact of a single new mouth to feed on the huge commons is only very slight, and is barely noticed. But the advantages to the shepherd of having the extra sheep in his flock are clear to all, and the other shepherds decide to increase their flocks as well. Each additional sheep adds only slightly to the common burden. But eventually, the collective pressure of adding more and more sheep is too much: The pasture becomes overgrazed and the field turns to dry dust. The scenario illustrates why maintaining cooperation among a large group is so difficult: The costs of a single "defection" is spread thinly throughout the entire group. Yet the payoff to the defector is quite high.

An analogous form of the "commons" problem is familiar to anyone who has gone out to dinner with a large group of people, with the agreement that they will divide the final bill equally among everyone at the table. Because the cost of each person's dinner is spread out among everybody, each person pays only a fraction of the extra cost of having an expensive meal such as a lobster or a big steak. Conversely, making the sacrifice of having an inexpensive meal such as a small salad doesn't help an individual who is strapped for cash very much, because the savings of the inexpensive meal are also split among everyone at the table. Thus in a group of ten people, for instance, someone who splurges on an expensive meal pays only a tenth of the extra cost, while the person who gets a cheap meal receives only a tenth of the benefit of the sacrifice. The result is that everybody orders the most expensive meal on the menu. This same problem occurs in other large groups as well—such as the U. S. Congress, where each senator and representative fights for special projects for his or her own state, but the payment for these projects is divided among the nation of taxpayers as a whole.

These kinds of cooperation dilemmas tax our evolved mental mechanisms to the fullest. Yet in many instances society does manage to avoid the "tragedy of the commons." In some cases, defections become classified as "crimes against the state." Like an insurance pool, a society bands together and collectively decides that defecting behavior will be punished by the group at large. Thus in cases where the enforcement may be dangerous, as in tracking down a murderer, or costly, as in a legal suit against the producer of a defective, harmful product, we have surrogates—police or government agencies—who act in our behalf. Society has less direct ways of punishing defections, too. Through language, custom, social pressure, and ostracism, humans have devised many ways of ensuring that the punishment for defecting on the group at large is high.

Societal standards of proper behavior are an important part of helping the group as a whole cooperate, says Axelrod. "They help ensure that most people will wait their turns in line at the supermarket, show up for appointments roughly on time, keep their lawns mowed," he says. "They discourage cheating on income tax returns, reinforce marital fidelity, and even, in the international arena, temper the behavior of whole nations." Without a "norm" in a society to back them up, laws often become all but unenforceable. The prohibition on alcohol in the 1930s, the ban against marijuana, and the fifty-five-mile-an-hour speed limit are all examples of laws that have been seriously eroded because of the lack of support from a corresponding social norm. On the other hand, many powerful social controls exist without needing the force of law: Slavery is no longer openly practiced anywhere around the world, not simply because there is legal prohibition against it, but because there is a powerful social prohibition against it.

In his computer simulation, Axelrod included a number of factors involved in maintaining norms in society. Suppose that a class of students is taking a test, for instance. Each student must weigh the benefits of cheating on the test against the cost of being punished if caught—what Axelrod calls the "temptation" factor. Likewise, a person who sees someone cheating must also weigh several considerations: One is the cost that the cheating might have on the group at large (the overall grade average will be affected, for instance); another factor is the personal cost to a player who punishes a cheater (having to "press charges," for instance, and the threat of retaliation). Axelrod also varied the "boldness" of the players—that is, how likely they were to risk getting caught—as well as the players' "vengefulness," that is, how zealous they were in punishing cheaters.

Throwing together these factors into his computerized societal stew, Axelrod let the culture "evolve" computationally by allowing the various strategies of different levels of boldness and vengefulness to interact, then increasing the number of strategies that were successful and decreasing those which were not. To his dismay, he found that a "no cheating" societal standard was very difficult to maintain. All of his simulations eventually evolved to the point where vengefulness was close to zero and bold cheating was rampant. Part of the problem is that the costs of enforcing the standards are high for individuals, while at the same time the benefits of their vengefulness are spread thinly over the population as a whole. In effect, says Axelrod, each individual faces the same question: "Why should I play sheriff if I'm not going to get paid for it?" he says. "The logic behind that question drives vengefulness down."

One way to turn this situation around is to include another factor in the simulation: punishing people who fail to punish cheaters. In real society, such tactics include ostracizing people who knowingly let their associates get away with reprehensible behavior. Politicians are taken to task for belonging to organizations that exclude minorities, for instance, even though they themselves might not endorse such a policy. Law-abiding people are expected not only to avoid associating with criminals but also to avoid people who associate with criminals. Likewise, most people would regard a person who knowingly let someone else get away with murder as morally culpable as well. "Not to join in the punishment," says Axelrod, "is itself taken as a betrayal of the group." When this additional level of "punish those who do not punish defectors" is added to the computer simulation, the cost of *not* punishing a defector becomes high, and the level of enforcement in the computerized society stayed high. The result was that a norm such as "It is wrong to cheat on a test" remained in force. "The entire system was self-policing," says Axelrod. "And the norm became well established."

Though this computerized society is obviously quite simplistic, the research has intriguing implications for everyday life. The work suggests that a society's values, conventions, morals, and standards of behavior are sensitive to people's willingness to punish those who break the social contract—and, more important, the willingness of people to punish those people who tolerate such defectors. Old-fashioned ideals such as community responsibility, good citizenship, and a person's reputation, character, and morality do in fact matter, even in a modern, cynical world: Leaders in business, politics, and society who merely wink at their associates who step outside the bounds of society's standards—or as seems more common today, feel it is sufficient to deny their own guilt and culpability—in effect lower the level of "social punishment" and cause society as a whole to suffer. When people continue to associate with neighbors or friends who cheat on their taxes, or sell cars, jewelry, and other goods to known drug dealers or other criminals—rationalizing that they themselves aren't committing a crime—the fabric of society is frayed further. The obstacle to maintaining values in our modern society is not simply the corrupt politician, shady banker, or remorseless street criminal—it is the fact that, despite their behavior, these people still have *friends*, and that these friends have friends. When people look the other way and "don't want to get involved," the enforcement that is necessary for cooperation often disappears, standards erode, and "defection" becomes rampant. "Erosion starts at the top and works down," says Axelrod. "Fifty years ago, a law-abiding citizen might well have avoided association not only with those who used drugs but also with those who associated with drug users. Today, though there may still be many people who have no close friends who use illicit drugs, it seems unlikely that many of us systematically shun people whose acquaintances use drugs."

Axelrod's simulation reveals that standards of behavior become stabilized only when the costs of defection start off high. In other words, norms must be present in the first place in order to persist in a society. Sometimes norms are created by people mimicking powerful social or political figures, such as copying the signature fashion of pop stars, or eating certain foods that are enjoyed by a powerful public figure, such as one President's love of jelly beans. In other cases norms are codified as religious and moral values in a society, and are often internalized so that not following them produces feelings of guilt and shame. The behavior of others is often a powerful influence on one's own behavior: People who would never think of throwing trash on the floor at an opera house might blithely leave their empty popcorn box on the movie theater floor.

Through mimicking people of power and influence, the cultural evolution of a cooperative behavior can even cause people to act beyond their most basic drives of self-preservation. One example of this is when Japanese pilots volunteered to fly suicidal Kamikaze missions during World War II: The Kamikazes weren't naive youths overcome with patriotism, but experienced pilots who were well acquainted with the realities of war. The growing ranks of pilots volunteering for death can be understood as an exaggerated evolution of a cultural principle of self-sacrifice for the sake of the country. In a way that's analogous to how sexual selection fuels the evolution of the elaborate tail of a peacock, the pressures of war caused the already established cultural trait of self-sacrifice among the Japanese to evolve into its most extreme manifestation, suicide.

CIVILIZATION'S TRIGGER

Given all the trouble that modern society causes for our Stone Age mind, why did our ancient ancestors suddenly end their hunting and gathering ways and settle down? What caused them to start huge construction projects, become specialists in various crafts, and stratify into state societies with a hierarchy of social, economic, and religious classes? There are two main trends of thought about the origin of modern states. The *conflict* school sees society arising as the result of a small group of people wresting control of key resources away from others and, by direct military force or indirect means, coercing others to labor for their benefit. This school, which includes many archaeologists who view society with a Marxist perspective, sees civilization as having arisen largely as a result of internal factors within the society: Fueled by the personal ambition of an elite, the state arises to benefit the ruling class, and all the institutions of the society, no matter how they appear to the casual observer, are ultimately designed to further the elite's interests.

In contrast to the conflict school of thought about civilization's origins is the *integrationist* school. In this approach, states are viewed as originating as a collective response to external forces that put a strain on a group of people living in an area. These forces can include food shortage, population pressure, climatic change, and competition or warfare from other groups, and can be made more pressing if the society is hemmed in by an ocean, mountain range, or territorial boundaries. A state may arise from a collective desire to solve problems, as in building irrigation canals, importing vital resources, or managing some other project that benefits all the members of society.

In fact, it is most likely that the cause of the rise of complex societies is neither conflict nor integration but both: A society-at-large needs to respond to a particular problem, and it gravitates toward those people whose personal ambition leads to their being given, or seizing, power. A state may also form for one reason and then change through time into something that serves some other purpose. Thus a state that forms for "integrationist" reasons may ultimately be overtaken by elites. Likewise, a state that forms through the ambitions of a ruling elite nevertheless may find itself striving to serve the needs of society at large in order for the elites to stay in power. In the end, both the conflict and the integrationist approaches to the origins of the state ignore the fundamental aspect of human psychology: cooperation. No matter how ruthless or benign a leader, there is a covenant between leader and led that works both ways.

A key part of creating complex societies is that rulers must maintain their *legitimacy* as players in a cooperative bargain with the "ruled." Whether power is seized by force or voluntarily given, rulers still must maintain the terms of their cooperative relationship with the people at large in order to get them to cooperate through paying taxes, serving in the army, etc. Legitimacy can be enforced at the point of a gun, but with greater shows of force come greater costs of maintaining the ruling infrastructure. If the costs of maintaining legitimacy become greater than the benefits given, the state may collapse. Thus states, like most interactions among humans all the way down to the one-to-one level, can be regarded as a cooperative relationship. Viewed as a whole, a state is a complex amalgam of competing interests and goals, with internal and external forces playing key roles in shaping how the society functions over time—and whether it survives.

The fact that the construction of irrigation canals, terraced farmland, and roadways ultimately benefits many people in the state—even if the elite get disproportionately more of that benefit—suggests that while elites may grab power solely for their own sake, they nevertheless must be responsive to the needs of their subjects as well. Machiavelli may have argued that "a prudent ruler ought not to keep faith when by so doing it would be against his interest," but such a ruler might have a hard time getting the populace to devote the kind of cooperation on which his regime might depend.

The necessity for leaders to have support from at least some of the population is evident even in societies of apes and monkeys. One might expect that the dominant male in a group of apes would be the strongest or most aggressive. But research shows that dominant primates are not necessarily those who are the most physically powerful, but rather those individuals who are most adept at *making alliances* among other members of the troop. Older males, for instance, have been shown to maintain their social position by craftily allying themselves with younger and stronger subordinates. "During a power struggle, male chimps will try to get support of females by playing with their infants," says primatologist Franz de Waal. "Presidential candidates do the same thing." Monkeys will also band together to topple a higher-up. In one experiment, a high-ranking female monkey was removed from her allies in one area and then placed in a separate area that was populated with a group of low-ranking females. The low-ranking females ganged up against her and reversed the hierarchy.

One clue to understanding the complex, cooperative relationship between the ruler and the ruled has come from archaeological research demonstrating that while the Inca rulers were never reluctant to use force to control their enormous empire, their real genius lay in their ability to cajole rather than coerce. The Inca recognized that maintaining an army in hostile country can be very expensive, and so preferred to use economic, political, and religious techniques to convince conquered tribes to live peacefully within the empire. Though the Inca are known to have committed bloody massacres and forced relocations of unruly tribes, in many cases the Inca left the existing social structure of a conquered people relatively intact. The local leaders often remained as governing officials in the Inca empire, though they answered to the Inca hierarchy. Peasants were obliged to pay a labor tax by working part-time in state-owned fields, serving in the army, or producing special goods such as pottery and cloth, but otherwise life for a typical farmer often remained largely the same. Studying daily life of a tribe known as the Wanka before and after their conquest by the Inca, archaeologists found that despite the burdens exacted by the state, the Wanka in many respects had a better standard of living than before they were part of the Inca empire, including less warfare with neighboring tribes and greater access to food and other goods. This is not to suggest that the Wanka were blissfully happy being subservient to their Inca rulers, despite these material benefits: When Pizarro began his efforts to topple the Inca government, the Wanka gladly joined him in the quest.

The Inca's governing strategy relied in part on their great skills at public relations. Archaeological excavations reveal that the Inca were masters at targeting their subjects' minds as much as their bodies. Excavating an Inca administrative center in a province far from the Inca capital of Cuzco, archaeologists discovered that one of the major functions of the site was not exerting direct political or economic control over the surrounding area but serving as a place to host lavish ceremonial feasts for the locals. These feasts were intended to create an ideological bond between the Inca rulers and the people of the province, a public relations tactic that helped reinforce the idea that being part of the Inca empire was more than merely working on state land or fighting in its army.

The dilemma the Inca faced in governing their subjects is still being grappled with by rulers today, who must constantly confirm their legitimacy to the ruled through economic benefits, ideology, and force. When the threat of force from the Soviet Union disappeared from Eastern Europe, for example, the ruling governments collapsed overnight because, without the backup of the Soviet Army, they lacked any ideological or economic legitimacy to continue governing. But during the surge of democratic demonstrations in China's Tiananmen Square, what was primarily an urban revolt was brutally shut down by troops whose soldiers were largely drawn from the countryside, not the cities. To people in the countryside the rulers had legitimacy, because farmers had reaped the economic benefits of the government's agrarian reform.

A COMPLEX PAST

The social complexity that characterizes modern societies has roots that stretch back as far as our species' "creative explosion" some 40,000 years ago. Body ornaments such as pierced carnivore teeth, carved pendants, and elaborate beadwork mark the beginnings of a complex social structure that arose through increased cooperation among people in a group, and among various groups in a region. Such sophisticated social complexity is evident in a site on the central Russian plains. Known as Mezhirich, the area holds the remains of five separate dwellings between twelve and twenty feet wide, each of which is estimated to have been the home of about fifty people some 15,000 years ago. Because wood was scarce in the chilly terrain, which was still in the grip of the Ice Age, the ancient people of Mezhirich made dwellings out of hundreds of stacked mammoth bones. It is estimated that each domelike hut would have taken ten people as many as five days to build. Though they did not raise crops, the people of the Russian plain apparently were surrounded by plentiful enough plants and animals to allow them to occupy these semipermanent structures some nine months a year, moving away in the summer to escape rising waters from a nearby river. Shells found at the site were traced to animals who lived in the Black Sea, which lay more than 300 miles to the south, suggesting that this ancient village was involved in widespread trading. Several deep pits had been dug into the earth near the dwellings, which apparently were used as deep freezes to store food. The fact that storage pits near some homes are larger than others suggests a social hierarchy in which some people had greater access to surplus food. Evidence of social complexity among our ancient ancestors also comes from several burials on the Russian plain that date back more than 20,000 years. The fossils in these graves are bedecked with thousands of ivory beads, weapons, and other finery that suggest a social stratification that enabled some individuals to accumulate jewelry and other goods. The fact that the nearby graves of two children are similarly bedecked with elaborate jewelry and weapons also suggests that this wealth and status was passed on through generations.

By 6,000 years ago, the transformation of our species' everyday life from its ancient roots in small bands of hunter-gatherers to the large, complex societies of the modern-day world was well on its way. Over the next two millennia large-scale city-states arose, the first of which were Sumerian in Mesopotamia. These were quickly followed by emerging city-states in the Nile Valley, the Indus Valley, Asia, Mesoamerica, and South America.

One of the major causes of the rise of these modern civilizations was the development of agriculture. For years, farming has been erroneously viewed as an "invention" and "revolution" that distinguished so-called advanced cultures from more primitive societies. Because we now live in an agricultural society, it perhaps seems logical to assume that once ancient people discovered that planting crops was possible, they would naturally begin farming, because it made their life so much better. A few researchers have gone so far as to suggest that agriculture—and the complex, sophisticated civilization that accompanied it—was spread around the world by people from these advanced cultures, who voyaged in boats to more "primitive" societies and passed the agricultural torch, Prometheus-like, to the supposedly "backward" peoples who lived there.

But, in fact, the rise of agriculture didn't come about as a result of one of our ancient ancestors suddenly saying, "Hey, let's invent farming," because to do so would have been well nigh impossible. Domesticated plants, like domesticated animals, are physically and genetically different from the wild versions from whence they arose. Many wild plants, for instance, do not "bloom" all at the same time, and so someone wanting to switch from hunting and gathering to farming couldn't simply start planting seeds and harvest them the next season. To make the kinds of biological changes in a plant that make agriculture possible, people must have been actively involved in the domestication process for many generations of the plant before they actually began farming.

In fact, the process of shifting a plant from a wild type to a domesticated type takes at least two decades of careful selection and breeding— nearly a generation in terms of human existence. So it is unlikely that our ancestors 10,000 years ago would have consciously undertaken a process that might have taken nearly a quarter of a century to achieve results—particularly if people were turning to plant foods because they were hungry. These "first farmers" would have had no way of knowing that such changes in a plant were possible in the first place, and no way of knowing that a potentially great payoff from such a process would ever come.

The birth of agriculture was not so much a revolution brought about by humans' "conquering" plants but a cooperative *coevolution* whereby both plants and humans changed their everyday way of life for their mutual benefit. Neither our ancestors nor the plants deliberately *tried* to change how each other lived, at first. But as our ancestors selectively gathered wild plant foods over the millennia, they became evolutionary forces that slowly altered the plants' ecosystem—creating evolutionary pressure for change along the way.

Even without conscious attempts at agriculture, for instance, our ancestors would have inadvertently influenced the success of a plant, through giving it protection, destroying competitors, and dispersing its seeds to new environments. For example, a tree that bore particularly large fruit might be spared being cut down for firewood or construction of a shelter, at the expense of other trees that were not as good producers of fruit. Our ancestors' impact on the plants' environment was perhaps more subtle and unconscious than the modern farming practices of tilling the soil or watering crops—but it had no less dramatic effect on the evolution of the plants. Through the disturbances of the environment brought about by their foraging habits, humans created opportunities for new ways of life for plants.

As a result, plants began to "compete" for the attention of humans. As our ancient ancestors harvested wild grains to eat, for instance, the few mutant strains of the wheat that had the least brittle fibers would have been picked first, because they were the easiest to harvest. These plants would have been the most likely to have made it back to our ancestors' base camp, and their seeds would have been dropped nearby. This would have resulted in a new stand of grain growing near the site, which in turn made it more likely to be harvested by the people living near there. As the process was repeated over and over, this mutant, more domesticated type of wheat would have become more prevalent in our ancestors' surrounding environment—even without our ancient ancestors' deliberately trying to make it so.

There are many cases of plant/animal partnerships in the natural kingdom. Some species of wasp serve as fertilization agents for fig trees, for instance, and squirrels serve as planters for acorns. Some birds serve as distributors of the seeds of the fruit trees they feed on. One of the most bizarre cases of cooperative relationships between a plant and an animal concerns a species of melon that can only grow after its seed has passed through the digestive tract of an aardvark. The aardvark eats the melon and afterward buries the seed of the melon along with its feces. A new melon grows from the seed, and the process is repeated again and again, as the aardvark in effect "farms" the melon.

Anyone who doubts that plants will join in a cooperative relationship with humans without people deliberately trying to enlist them need only look as far as their backyard garden: There, the persistent presence of weeds testifies that plants need no special attention from humans to become part of our everyday life—weeds have adapted themselves to the new environments created by humans even though people are consciously trying to keep the plants out. The difference between crops and weeds, in other words, is simply that we don't have any use for weeds—but both categories of plants have been "domesticated" over the millennia by human activity.

MAN'S BEST FRIENDS

The idea of domestication being a process of cooperation and coevolution between species, rather than conquest, applies to animals as well. Domesticated dogs, which first begin to appear in the fossil record some 12,000 years ago, are physically and behaviorally different from the Asian wolves from whence they are descended. Compared with wolves, dogs have shorter muzzles, mature faster, and are smaller overall—a package of traits that represents an increasing trend toward "juvenilization," that is, retaining puppylike characteristics into adulthood. As with plants, these physical and mental changes must have taken many generations to evolve, and so dogs must have had a long association with humans before people deliberately began to breed the animals.

The juvenile traits of wolves that persisted in adult dogs extend to behavior as well: Adult dogs are less defensive than wolves and will engage in puppylike "soliciting" behavior such as seeking affection or begging for food—behaviors that disappear in wolves when they reach adulthood. Because people have evolved to react kindly toward these kinds of juvenile behaviors in their own children, the wolflike ancestors to dogs who displayed such juvenile traits endeared themselves to humans and so were more likely to have benefited from the human/dog relationship. These wolves in turn produced more offspring, perpetuating the trend and eventually resulting in a new species of animal—the dog—that adapted itself to the new social "environment" of humans. The highly social nature of wolf society made wolves prime candidates for a partnership with humans: Besides hunting in groups and sharing food, wolves live in a hierarchical society dominated by a pack leader—a role that human dog owners take on today.

The cooperative association between the wolf-dogs and our ancient ancestors was as beneficial for one as the other: The dogs received protection, warmth, and food in exchange for their help in hunting and tracking. Domesticated dogs appear to have been selected to help serve as sentries as well—wild dogs and wolves don't bark. Other animals such as horses, cattle, sheep, and chickens also appear to have become domesticated not merely because of the human desire to use them for food and transportation but also because of characteristics in the animals that lent themselves to entering into such a partnership with humans. Corralling and controlling the creatures that eventually became our modern domesticated animals would not have been easy for our ancient ancestors, who lacked the technology to build fences. Domestication of these creatures relied in part on these animals' willingness to go along with the bargain. One big reason we eat beef rather than zebra meat is not because beef is tastier or more nutritious, but simply because cattle will live with people and zebras won't.

Though the benefits to cattle of domestication might appear dubious to anyone who has visited a stockyard, it is nevertheless true that without having coevolved as part of the cooperative relationship with humans, all these animals and their ancestors would have become extinct long ago, just like the aurochs and mammoths depicted in our ancient ancestors' cave paintings. Likewise, the population of the majestic African elephant is dwindling and threatened by extinction because of poachers who kill them for their ivory; Asian elephants, on the other hand, are "domesticated" and used in agriculture and industry and so are not in danger of becoming extinct. Some people characterize the domestication of animals as tantamount to slavery. But the evolution of domesticated animals demonstrates that, unlike the practice of slavery, these animals are not merely captives kept against their will, but new species that have adapted themselves to the "environment" of humans. If everyone on Earth suddenly decided to become vegetarians, as well as freeing their pets, then the populations of cats, dogs, chickens, cattle, sheep, pigs, goats, and horses in the world would rapidly dwindle—the only way these animals can survive is as partners with people.

HISTORICAL ACCIDENTS

The shift to farming actually represented a more profound social change than the mere creation of the "new" technology of agriculture. After all, there was already a longstanding, cooperative relationship between plants, animals, and our ancient ancestors before people finally shifted to deliberately farming the soil and herding. This dramatic social change could only have come as a result of the convergence of several factors, including a growing population and a changing climate. And because people rarely give up one way of life for another that is radically different, it is very likely that our ancient ancestors would not have made the drastic change to farming unless they were forced to by circumstances beyond their control.

Farming first appears around the world at a time that roughly coincides with the radical shift to a warmer global climate that occurred as Earth's most recent period of intense glaciation ended. For nearly 100,000 years, Earth's northern regions had been covered in mile-thick glaciers that reached down to what is now Chicago and Boston, and covered much of Europe and Asia as well. The spreading of the glaciers was part of an ongoing cycle of the waxing and waning of Earth's ice sheets that first began during the times of *Homo habilis* some 2 million years ago. In effect, the "Ice Age" has been ongoing since the dawn of the human species. Over the millennia, Earth's climate has swung between long periods of intense glaciation, lasting about 100,000 years, followed by periods of about 10,000 years of balmier climes. Earth shifted out of an icy climate into a short period of warmth about 125,000 years ago—about the time that the first *Homo sapiens* roamed the earth. Then it quickly got colder again, and remained so for most of the time that our species has been in existence. The last time the shift to warmer temperatures occurred was about 10,000 years ago, and nearly all things we associate with our modern everyday life—farming, metals, writing, civilization, and so forth—appeared during this relatively brief warming spell. In fact, according to the pattern of the last 2 million years, Earth is due for another period of spreading glaciers and cold temperatures sometime soon—though the potential climatic effects of our dumping carbon dioxide and other so-called greenhouse gases into the atmosphere may interfere with this change.

The glaciers reached their fullest extent about 18,000 years ago and began to recede again, and as they pulled back, many of the species of large animals that roamed the earth began to disappear, victims of the changing environment and of being hunted to extinction by *Homo sapiens*. With much of the larger game disappearing around the world and the human population growing, people began to expand the range of resources they used to survive, shifting from big game hunting to trapping small animals, birds, and catching fish and other marine animals. With fewer migrating herds of animals to rely on for food, our ancient ancestors turned to eating plants to survive, which resulted in increasing pressure to become more sedentary. Since people couldn't take the wild plants with them, they tended to stay near where the plants grew.

One place where the sedentary life flourished is in the Levant, an area in the Middle East along the Mediterranean Sea and stretching northward, including the modern political states of Israel, Syria, Jordan, and Lebanon. From about 12,500 to 10,500 years ago, this area was populated with people who are known as the Natufian culture. The Natufians became the first people to become permanently settled, and it came about as they were forced to respond to a one-two environmental punch that came as the earth underwent its massive bout of global warming. As the glaciers receded, sea levels rose nearly 400 feet in some places, cutting off Alaska from Siberia and New Guinea from Australia. The Arctic-like tundra that had covered much of Europe, Asia, and North America in the preceding millennia was replaced by forest and grass, and the world's climate rapidly warmed to one that is much like today's. This climatic change was acutely felt in the Levant, as the lush hill-country filled with a bounty of plants, and the animals were hemmed in by the mountains to the north, the Mediterranean Sea to the west, and spreading deserts to the south and east. The Natufians responded to their ever diminishing range by abandoning their wandering ways and settling down, though they continued to hunt and gather their food to survive.

Evidence that the Natufians settled down comes from examining the fossilized teeth of gazelles, which grow faster or slower depending on the season, producing a distinctive pattern on their teeth. The remains of sites left by the Natufians' immediate predecessors in the area contain only teeth from gazelles killed in a single season, such as the spring or autumn—suggesting that these camps were temporary places early humans occupied for only part of the year. At the Natufian sites, however, there are teeth from animals killed at all times of the year, indicating that the Natufians had settled in one spot more or less permanently and were hunting the gazelle around them. It is also about this time that fossils of the modern version of the "house mouse" begin to appear at the sites as well. Besides hunting local animals, the Natufians appear to have depended on harvesting, processing, and eating wild plants such as wheat that grew in the area. Archaeologists have found grinding stones and sickles, and Natufian fossils bear evidence of worn and pitted teeth that would have arisen from eating tough plant fiber. This particular combination of social complexity, sedentary life, and technology for harvesting, processing, and storing wild plant foods made the Natufians prime candidates for being the first farmers.

The Natufians survived in their new life as sedentary hunter-gatherers until, about 11,000 years ago, they were hit with the second part of the climatic one-two punch: a sudden cold snap. For reasons that are still unclear, the slowly warming earth appears to have suddenly reversed itself and shifted back to a glacierlike climate for about a thousand years. The cool weather once again caused the Levant to undergo a radical shift, as plants that had been adapted to the warm Mediterranean climate disappeared, and many of the animals that fed on them followed. It must have been a wrenching time for the Natufian people. After generations of warming climate, the wind suddenly turned cold, the summers shorter, and the ground more parched. What was once a land of bounty became grudging in its sustenance. Having already adopted a sedentary life for generations, the Natufians were faced with a choice: Start moving again or find a new way to survive. They turned to agriculture.

The shift to farming required no grand technological breakthroughs: Despite their hunting and gathering life-style, the ancient people of the Levant had been familiar with cereals and grains for generations. At one archaeological site in Israel, for example, scientists unearthed grains of barley that date back 19,000 years. The Natufians' long association with grains suggests that taking the step of deliberately planting seeds for the next growing season would have appeared as an obvious move, particularly to people who were searching for ways to extend their food resources.

Of course, it was not necessary for other peoples around the world to undergo such severe climate change to turn to agriculture for survival. Merely by becoming sedentary, people set the stage for farming. Staying in one place tends to result in a boom in population: Because women aren't on the move all the time, they can have another child while a previous child is still semi-helpless. This ability raises the birth rate, which leads to a population growth that in many societies would have eventually outstripped the available resources around them.

FARMING'S MIXED BLESSING

Far from a glorious ascent toward civilization, as it has long been depicted, our ancestors' shift to the farming way of life resulted in many new hardships. The switch in diet from that of a hunter-gatherer to that of a farmer resulted in bouts of malnutrition and chronic disease. Because they were constantly on the move, hunter-gatherers' everyday diet included many different kinds of foods obtained from a wide variety of sources, making it more likely that vital minerals and other nutrients that were lacking in the food at one spot were present in the food found at another place. The migrating ways of our hunter-gatherer ancestors also kept many diseases to a minimum, because people in effect outran the organisms that preyed on them.

In contrast, the shift to farming—and the sedentary ways that came with it—introduced a whole new set of problems for our ancestors. The plants and animals that became domesticated were selected as much for their ease of harvesting, herding, and storage as for their taste and nutritional value—a practice that continues today with the modern, base-ball-like tomato, for instance—and people's reliance on only one or two crops sometimes resulted in vitamin and mineral deficiencies. People living in close proximity all year round created new opportunities for the spread of diseases—including the new diseases of smallpox and measles, which originated as diseases of domesticated livestock that adapted to human hosts. Sanitation and waste disposal could no longer be achieved merely by moving somewhere else.

These consequences suggest that the dramatic shift in how our ancestors went about their daily lives was a desperate act for survival, not a glorious invention. Small wonder that one of the most powerful cultural icons to come out of the Levant during this period—the biblical book of Genesis—describes humans living in a lush garden of plenty, only to be banished to a life where people had to eat "by the sweat of thy brow," tilling the land. The similarities in the story of Adam and Eve's banish-ment out of the Garden of Eden to our ancestors' shift from the hunting-gathering way of life to the hardship of farming is perhaps more than coincidental. The switch to agriculture was a critical turning point in our species' history. It resulted in by far the biggest man-made "disaster" ever foisted on Earth's environment—agriculture—and caused sweep-ing changes in the kinds of plants and animals that live on the planet.

Yet if farming introduced a plague of new problems for our ancestors, it did provide a crucial advantage that inevitably led to its spread around the globe: The simple fact is, farming the soil produces more food per acre than hunting and gathering, and therefore more people could live off the resources in a given area of land. By making possible a huge pop-ulation boom, farming became a necessity: With more mouths to feed, there was no way our ancestors could return to hunting and gathering as a way of life. Once they began to farm there was no turning back—a sit-uation that prevails today. Farming may have been a last-ditch effort by our ancestors to survive, but it created changes they could never have foreseen—it allowed a population boom in our species, which led to our occupying nearly every area on Earth and enabling the creation of cities, states, and nations.

• • •

FORCED CIVILIZATION?

The complexities and potential pitfalls of congregating into large groups suggest that, as was the case of the adoption of agriculture, people gathered into large groups as much out of necessity as choice. The pressure to shift from living as part of a small group of hunter-gatherers to living in a large, stratified society may have come not just from the surrounding climate but as a result of the presence of other large-scale societies. With their power to perform large, difficult tasks, complex societies pose formidable challenges to smaller groups around them. Cooperation is not only useful, it is very powerful, and many challenges that would be impossible for individuals to overcome on their own can be performed by people working together. Large-scale building projects such as irrigation canals are a hallmark of complex societies—but so are powerful armies. With its ability to mobilize groups of people, efficiently process information, and perform large-scale tasks, a complex society becomes more than a mere collection of people. The state, in other words, becomes an entity in its own right, undertaking certain projects, such as waging war, that may be ultimately "good" for the state but bad for various individuals in the state.

To an increasing extent through most of recent human history, one of the largest challenges to any particular group of people was *other* groups of people. These challenges need not have been merely the obvious actions of threatening war or overtaking some important resource, for example. States, like individuals, also engage in cooperative interactions such as trade, mutual protection, or resource sharing. As with individuals, the outcome of these exchanges can be greatly influenced by how each partner judges the other's feeling about the value of the particular item of exchange—who most needs a barrel of oil, for instance—as well as the ability to punish if one or the other "defects." In these kinds of cooperative transactions, large, highly organized groups of people have a great deal of leverage over smaller, less organized groups or individuals.

Thus one powerful reason people congregated in large groups is that they didn't have much choice: Once one group of people became organized into a complex society, their cooperative power posed such a challenge to smaller groups that to *not belong* to a large group would result in more disadvantages than any disadvantages that resulted from *being in* the group. In other words, since people couldn't beat them, they had to join them. As evidenced by the shrinking populations of hunter-gatherers worldwide—as well as the devastation that occurred when people from European states encountered the hunter-gatherers living in North America five centuries ago—states have the power, for better or worse, to push less complex forms of social organization off the map.

While people tend to equate civilization with progress, civilization is at least as much cause of trouble for the human condition as relief. Population growth and death rates and signs of physical stress from disease or malnutrition in skeletons of various ancient peoples suggest that the rise of civilization resulted in a net *loss* of general well-being for most of the people who lived in "civilization" throughout history. Though a small elite in eighteenth-century Europe lived better than their predecessors, for instance, for the vast majority of people in cities and states life expectancy was less than that of ancient hunter-gatherers.

These data suggest the philosophical question, *Is civilization all that good for us?* Regardless of how a person might answer that question—and one's answer is as much a product of personal philosophy as scientific data—the query points out the fallacy of simply assuming that the evolution of human society is an inexorable march of progress toward a better life, fueled by the human capacity for innovation and technological breakthroughs. In fact, one could argue from the opposite perspective: That the series of "revolutions" in agriculture, technology, and the rise of the state are but a chain of desperate measures undertaken by our ancestors to survive in a world of increasingly dwindling resources and a population that continues to grow beyond its means. Ultimately, however, both perspectives suffer from the assumption that modern civilization represents some sort of endpoint in the evolution of human society. In one respect, it is: There is no way that the world's 5 billion people could return to hunting and gathering or even small-scale agriculture and survive. There are simply not enough resources or space.

Yet simple comparisons of present and past obscure the fact that human society has never stayed the same, but rather has always served as a means for changing itself. People in eighteenth-century England may have been no better off than Cro-Magnon hunter-gatherers of Ice Age Europe, but twentieth century people, on average, are far better fed and live longer than at any time in human history—and this condition is the direct result of advances in medicine, sanitation, and agriculture that could only have come about through the auspices of civilization, including those that flourished during the eighteenth century. Furthermore, not only has society changed but the criteria on which it is judged *good* or *bad* have changed, too. Health is but one factor to consider in judging the quality of life: The Wanka tribe of fifteenth century South America may have been better fed and freer from violence while under the yoke of the Inca, but whether they were better off is a matter of personal taste—*they* apparently didn't think so.

Relative latecomers in the human saga, complex societies have taken on a dazzling variety of forms since groups of people first began staying in one place some 10,000 years ago. Whether modern human society as it now stands will continue to be a viable strategy for survival into the next thousand years, or even through the next century, has for years been a rich vein for professional prognosticators and no doubt will continue to be. Yet if the past is any indication, the human species has by no means exhausted the possibilities of the kinds of worlds it creates for itself to live in. The question is, will our consummate skills as cooperating creatures eventually be the cause of our demise?

Epilogue

US VERSUS

THEM

Humanity is just a work in progress.

—TENNESSEE WILLIAMS

Scientists and lay people alike have long viewed the evolution of the human species as one of the greatest dramas in the unfolding of life on Earth—a tale of passion, challenge, and high adventure, with everyone's favorite character, *themselves,* in the lead role. In these scenarios, our ancient ancestors are portrayed as lone heroes who leave the safety of their forest home, abandoning their less ambitious brothers, the apes, to face alone the enormous challenges of a new world of the savanna. They overcome these great challenges with a special gift, their large brain. And in the end, this great gift also becomes their greatest challenge: Their intelligence has given them the power to rule the earth, but with that power comes the potential for self-destruction. "They find they have the power of the gods," says anthropologist Misia Landau, who studies these "origins" tales, "but not their wisdom."

The tendency to see our ancient ancestors as the heroes of an evolutionary quest is perhaps understandable. A crucial component of our evolved minds is the ability to interpret the behavior of other people, animals—even the weather—in terms of their goals and intentions, as in "He only did that because he's angry," or "It's raining because I planned a picnic today." So it comes as no surprise, perhaps, that we mistakenly regard evolution as having goals and purposes, too: Specifically, we tend to think that evolution's ultimate goal and purpose was to produce *us*. This vision of ourselves as the culmination of billions of years of the evolution of life on Earth is aptly illustrated in the classic diagram that has graced schoolroom walls and textbooks for decades. The diagram shows a primitive fish climbing out of the sea, followed by a succession of animals ascending a hill, from something that looks like a hedgehog to an ape to an ancient-looking hominid to, at the very top of the summit—and capping off the whole process, it is implied—a modern human, the icing on the evolutionary cake.

But evolution does not plan for the future; who we are today merely reflects the kinds of strategies that were successful among our ancient ancestors in the past. And while overall, the evolution of all life on Earth might be characterized as increasing in complexity, it is meaningless to say that one animal is "more evolved" than another. Contrary to popular misconception, we did not "evolve from chimps"; rather, chimpanzees have evolved just as much as we have over the past 5 million years since our two lineages split from a common, apelike ancestor. The human species is best characterized not at the summit of a hill of ascending animals, but merely as one branch of a low, fat bush that has many branches spreading out in all directions. In fact, many of the ancient, now extinct branches of the human "bush" extend far longer than our own.

As with the evolution of any species, the emergence of modern humans was an isolated, unique event, no more predestined, inevitable, or predictable than the appearance of wombats or houseflies. We are simply one highly successful example of the more than *half-dozen* brainy, two-legged primates that peopled the earth over the past 4 million years; indeed, we are but one of several humanlike kinds of primates who were sharing the planet as recently as 40,000 years ago. The fact that our species is the only such creature remaining in modern times testifies to our remarkable adaptability—and good fortune, perhaps—but not to our triumph as an evolutionary experiment. The puny, pea-brained, tree-climbing *Homo habilis* may not seem like much of a match for *Homo sapiens*, but it walked the earth for more than a million years—ten times longer than modern humans have existed so far. To equal *H. habilis'* tenure on Earth, modern civilization would have to survive not just through the twenty-first century, but for another *9,000 centuries* after that.

INVENTING THE FUTURE

Ironically, if we are going to surpass the tenure of our ancestor species on Earth, we will have to overcome the same fundamental challenge that they faced on the savanna eons ago: *each other.* It is not the threat of nuclear annihilation, the prospect of vast climate change, or the spread of deadly diseases that endanger our existence in the future, but the very thing that brought our species into the twentieth century in the first place—our evolved psychology, with its amazing abilities to cooperate. Our ancestors evolved brains that were custom-designed for helping them get along with each other, cooperating to do together what would be difficult, if not impossible, to do alone.

Yet this same evolved, Stone Age mind poses our greatest challenge. We live in unprecedented times, evolutionarily speaking: For nearly all of the human saga, a time period stretching back millions of years, our ancestors lived in small groups that, while often linked to other groups in vast trade networks, lived their everyday lives more or less isolated from each other. The result is that the modern human brain has been designed by evolution for a social life that is not very different from that of the Cro-Magnon who lived in Europe some 30,000 years ago. Indeed, even though we now live in large, modern societies of millions of people, the number of people we know *really* well—about 100, as opposed to 500 or 5,000—hasn't changed since the time of our ancient ancestors. Our evolved psychology simply can't deal with the complexities of forging cooperative relationships with many more than a small group of people, because cooperation is extremely dangerous, extremely rewarding, and extremely taxing on the brain.

When the size of the brain of various other primate species is compared to the size of the social groups those primates typically live in, a simple relationship appears: The larger the brain, the larger the group. If this brain/group size relationship is then applied to humans, the ideal group size is predicted to be about 150 people. Intriguingly, this is close to the size found in many hunter-gatherer clans and the first villages thousands of years ago. This group size turns up in modern industrialized societies as well: The basic fighting unit of a modern army is the company, which consists of about 130–150 soldiers; a standard practice in business is to shift to a hierarchical chain of command in a company that grows larger than about 150. Even academics seem to cluster into groups of about 100–200 scholars per discipline. A group of religious fundamentalists known as the Hutterites limits its group size to 150 people. The reason they give for doing so gets to the essence of the evolution of the human psyche: When their groups are bigger than that, say the Hutterites, individuals can no longer be controlled by social pressure alone.

The challenges faced by our Stone Age mind amidst the new social environment of huge cities and nations suggest that our species may not be immune to further evolutionary pressures, and so may not have finished its evolutionary journey, as many people assume. Technology has drastically reduced the evolutionary pressures on our body, but our new social surroundings still put a premium on having a mind that can get along with others. Similar pressures resulted in a host of evolved mental mechanisms in our ancient ancestors—and there is no reason to suppose that this evolutionary pressure has ceased. But it took millions of years to evolve the mind within our skulls today, and our species has been living in its radically new social surroundings only a few thousand years— far too short a time for any new evolved mental mechanism to arise.

The human species has succeeded beautifully at being able to cooperate as individuals in small groups, through language, culture, and other unique facets of our minds that help us negotiate the tricky business of getting along. But within the vastly different environment of complex societies, we now find ourselves dealing with far more people, many of whom have different goals or are fleeting strangers—awash in a sophisticated technological environment where it is sometimes hard to keep track of who's who and what's what. Our food no longer comes from a neighbor whom we see day after day, but from an anonymous member of an anonymous chain of grocery stores. Our personal protection stems not from the threat of retaliation by our families and close friends, but from a trained elite whom we've never met. We are surrounded by people who worship different gods, harbor different attitudes about what kinds of behavior is appropriate or inappropriate, and disagree about fundamental issues concerning life and death. Through it all, we manage to knit together the fabric of our communities, through trust, tolerance, and a common belief in a shared future—but our cooperative skills are stretched to their utmost.

GROUP AGAINST GROUP

Sadly, it is these very skills that allow us to bond together in coalitions that also enable us to form groups *within* groups. People use their superb cooperative skills to forge powerful coalitions, cartels, interest groups, gangs, monopolies, and cabals. Against these groups, individuals are sometimes at a tremendous disadvantage, as are other, less tightly knit groups. The very existence of one group of cooperators creates a strong pressure for others to join, too, or form their own group. The result is that our species' supreme abilities at joining into cooperative groups, which enabled our ancestors to survive over the millennia, can also backfire into xenophobia, rampant nationalism, class warfare, and racism today.

It is precisely our evolved, cooperative human psyche that has fostered some of the greatest cruelties in modern society. Faced with a sea of anonymous faces as societies grow beyond the 150 or so individuals we can keep track of in our day-to-day social bargaining, people often resort to using simplistic labels—economic, class, or race—to distinguish potential cooperators from defectors. Unfortunately, a situation where one group is antagonistic toward another can persist indefinitely in a society, despite the long-term problems it causes for the society as a whole. In a variation of his computer tournament for cooperation, Robert Axelrod modeled how such an antagonistic standoff between groups of people can arise when people resort to using labels to decide what kind of strategy to play with each other all the time.

For example, suppose each person in a society wore a hat that was either blue or green. Suppose further that each person used an identical strategy when interacting with another person. The strategy is, *When in-*

teracting with people who are wearing a hat that is the same color as yours, play TIT FOR TAT; *if the hat is a different color, always defect.* That is, if a green hat meets another green hat, the two start out with a mutual co-operation—as dictated by TIT FOR TAT—and reap the mutual benefits, and likewise if a blue hat meets a blue hat. If a green hat meets a blue hat, however, each treats the other as a potential defector and both immediately defect.

To his dismay, Axelrod found that his computer model showed that a population consisting of groups of people wearing blue hats and green hats and playing this strategy is "evolutionarily stable"—that is, once the whole population is playing this strategy, no other person playing a different strategy can do better. If a blue hat bucks the system and tries to cooperate with a green hat, for instance, he or she will get defected upon in return. The tragic result is that both blue- and green-hatted people are stuck defecting on each other, and so do worse overall than if everybody simply ignored which hat the other was wearing and played TIT FOR TAT with each other.

More troublesome, from the point of view of relations among the various ethnic and special interest groups that exist in modern society, is the finding that if one of the groups is smaller than the other, each person in that "minority" group will do worse overall than those in the majority group. One simple reason is that if the two groups interact randomly, people in the majority group will interact with each other more often, simply because there are more of them, and so they will more often reap the benefits of mutual cooperation. Conversely, the people in the minority will interact more often with those in the majority, with the resulting punishment of mutual defection. In addition, since there are more people in one group than in the other, mutual defections between members of the two different groups count less for the larger group as a whole, because the cost of the defection is spread out among more people. The findings have obvious implications for race relations—after all, if members of two different races come to an interaction mutually suspicious of each other, the result is the self-fulfilling prophecy of mutual defection.

In fact, however, racism is not part of our evolved psychology. Rather, our Stone Age mind is tripping over something that is more fundamental: "us" versus "them." Just as our propensity to eat fatty foods stems from the scarcity of fat in the past, our willingness to regard others as "them" has roots in the societies of our ancient ancestors. Early humans spent most of their lives interacting with very few people, all of whom looked and talked just as they did—and for whom strangers and members of neighboring groups were "scarce." When the "shadow of the future" is small—that is, when it is unlikely you will ever meet that person again—it is unlikely that cooperation will flourish. Thus our evolved psyches regard anyone who is not part of our intimate group as a "them."

Skin color is simply one of the more obvious labels that distinguish people from each other and make it easy to lump a stranger in with a set of preconceptions about how people in such a group behave, with sadly predictable results. But skin color need not be the only obvious marker that separates "us" from "them." Accent, language, ethnicity, stature, hair color, and clothing are all used on every continent to distinguish one group from another. Group-against-group conflict takes place throughout human society, with rich versus poor, liberal versus conservative, Catholic versus Protestant, East Coast versus West Coast, Celtics fans versus Lakers fans, hipsters versus "squares," and so on. Even groups that seem cohesive when confronting other groups are often splintered into smaller subgroups within themselves, such as the distinctions between "old" rich and "new" rich, Marxist Leninists and Marxist Maoists, lighter-skinned African Americans and darker-skinned African Americans—even aficionados of the "acoustic" Bob Dylan and fans of the "electric" Bob Dylan.

More troubling than our capacity for bigotry and racism is that it stems from a more sinister attribute: Our willingness to denigrate *any* group that exists outside our own. This willingness to deem others as "them," and to regard them as less worthy than ourselves and other members of our own group pose the biggest challenge for the human species.

The inclination to form bands, cliques, clubs, secret societies, and "in" groups to the benefit of themselves and the exclusion of others is part of the coalitional psychology that enabled our ancestors to thrive. Recent psychological experiments suggest that the drive to identify with the interests of one's own group can knit together even total strangers for an afternoon. In one series of experiments, a group of nine volunteers who did not know each other were each given five dollars and asked to make a choice: If five or more people gave up their five dollars to be distributed among the whole group, everyone in the group would receive an additional ten dollar bonus. Those who didn't contribute their money nevertheless shared in getting the ten dollar bonus without giving up anything. On the other hand, if fewer than five contributed, no bonuses were given—and the few who contributed their share lost their money. When the people in each group were allowed to interact for even just ten minutes, they earned the bonus every time, typically with seven or eight people contributing. Those groups where members were not allowed to interact fared less well, achieving the bonus only 60 percent of the time. While these results might make one hopeful about the cooperative nature of the human mind, a follow-up experiment revealed a disturbing wrinkle: When the people in the group were told that their contributions and bonuses would go to help not themselves but would be given to people in *another* group, the willingness of the members of a group to contribute their share for the "greater good" all but disappeared. It appears that the urge to better the overall outcome for one's own group is what elicited the cooperative behavior among the group's members—and not simply a desire to be altruistic toward others.

A GLOBAL FREE-FOR-ALL

In contrast to the lives of our ancient ancestors, the potential for group-against-group competition adds a new dimension to the cooperative process in large societies. The challenges for individuals in a cooperative relationship include recognizing one's partner, keeping track of past exchanges in the relationship, assessing costs and benefits, catching cheaters, and, most important, seeing into the future. All these challenges seem easy for us because the human mind has had millions of years to evolve for the task. Our species is living, glorious proof that the "cooperation dilemma" of everyday life can be overcome, with tremendous rewards.

It remains to be seen, however, whether *groups* of individuals can overcome the cooperation dilemmas that now exist on a global scale. The world is one big arena filled with cooperative coalitions of people that have varying goals and agendas. From racial groups to religious factions to international businesses to nations to multinational alliances, these groups are continually engaged in a cooperation "dilemma" of global proportions, under conditions of near anarchy.

The fact that the human mind bears an evolutionary legacy forged during the times of our ancient ancestors does not suggest that racism, xenophobia, and ethnic conflict are *inevitable;* the new research into evolutionary psychology suggests that our minds have evolved the mental tools that may well be up to the challenges of the future. The key to our species' success is our great skill in making close alliances with others, encouraging cooperative relationships with even the most self-centered people, and, most important, recognizing that, just as it was in the times of our ancestors, the future depends on compromise and mutual getting along. The research into our Stone Age mind reveals that being cooperative isn't merely the same thing as being nice. Our greatest skill is forging cooperative bonds with just about anybody—even old enemies: As history has shown, one group's worst rival can within a short time become its closest ally.

Like our ancient ancestors, the various groups in the world today can rely on themselves and each other to share mutual benefits and punish defectors. As with our ancestors, fostering cooperation among today's various groups will depend on the groups' abilities to forge new cultural inventions, communicate with each other, and have the leadership to overcome the short-term allure of defecting for the long-term benefit of cooperating. The nations of the world need no "outside" threat—such as the oft-evoked invasion by space aliens—to bind them in cooperative bonds. Ultimately, a world that is now characterized by "us versus them" will truly begin to get along not when faced with alien invaders or when under the yoke of some worldwide central government that banishes by decree prejudice and racism, but when the various groups, large and small, realize that they have an inevitable, common future.

The challenge of multiple groups' getting along is unprecedented in our species' tenure on Earth. Large-scale society has existed for only about 5,000 years—a tiny fraction of the human saga. For most of human history the standard way of life for our species was to grow as a group, split, and move to another place if there was trouble. Now there is no other part of the world to go to. Both geographically and in terms of Earth's resources, there is no new part of the planet left untouched by humans. It may be only now, in this most recent phase of our species' evolution, when we have finally filled every nook and cranny on Earth, that the various groups around the globe will begin to recognize that their future is intimately connected with each other. How our evolved mental mechanisms help us cope with being cooperative members of groups, who in turn must live cheek-by-jowl with other groups, is a crucial part of understanding our past, just as it is a vital part of understanding where society might go in the future.

The human saga is a dynamic, ever changing process. The new research into our deeply social intelligence reveals that our ancient ancestors never lived in stable, perfect harmony with Nature or each other, and there is no such thing as an ideal, natural human way of life. Instead, our species has been constantly negotiating its place in Nature, reinventing itself and its behavior to adapt to an ever changing world. We continue to face new challenges, brought about by the technology of war, the potential for climate change, and the pressures of a growing population. Yet in some respects these challenges are no less daunting than those faced by our ancient ancestors on the savanna millions of years ago. They survived by inventing the future. Our task is nothing less.

Notes

INTRODUCTION

19 Paleoanthropologists have long depicted our ancient human ancestors as a kind of Robinson Crusoe: See Nicholas K. Humphrey, "The Social Function of Intellect," in *Machiavellian Intelligence: Social Expertise and the Evolution of Intellect in Monkeys, Apes, and Humans,* ed. Richard Byrne and Andrew Whiten, Oxford University Press, 1988.

20 There have been no less than six different kinds of upright-walking, intelligent, apelike creatures that have walked the earth over the past 3 million years: For an excellent, balanced academic overview of human evolution, see Richard Klein, *The Human Career: Human Biological and Cultural Origins,* University of Chicago Press, 1989.

20 It doesn't take many smarts to escape a tiger, find a safe place to sleep, or pull fruit from a tree: See S. Pinker and P. Bloom, "Natural Language and Natural Selection," *Behavioral and Brain Sciences,* vol. 13, 1990

21 All help create a social hotline that helps us determine who to cozy up to, who to butter up, and who to avoid, without having to experience it firsthand: See John Tooby and Irven DeVore, "The Reconstruction and Hominid Behavioral Evolution Through Strategic Modeling," In *The Evolution of Human Behavior: Primate Models,* ed. Warren Kinzey, State University of New York, 1987.

23 The scientific stories we tell of who we are and how we came to be are more than mere information, for they serve as cautionary tales that remind us of our place as creatures of Nature: See Matt Cartmill, "Four Legs Good, Two Legs Bad," *Natural History,* vol. 11, 1983.

STORMING THE CITADEL

25 "Our struggle is to figure out how biology affects us, not whether it does."

Stephen Jay Gould, quoted in Carl N. Degler, *In Search of Human Nature: The Decline and Revival of Darwinism in American Social Thought,* Oxford University Press, 1991.

25 "Anthropologists have always assumed that evolution carried the human species up to the dawn of modern society and then left us there": See David Pines, *Emerging Syntheses in Science,* Addison-Wesley, 1988.

26 What is the mind for?: See "From Evolution to Behavior: Evolutionary Psychology as the Missing Link," in *The Latest on the Best: Essays on Evolution and Optimality,* John Dupre, ed., MIT Press, 1987.

27 Prominent scholars sought to protect the purity of their races from genetic "mongrelization": See Robert Richards, *Darwin and the Emergence of Evolutionary Theories of Mind and Behavior,* University of Chicago Press, 1987.

27 Early in this century, a number of state laws in the United States authorized the sterilization of "undesirables" because, as one California statute puts it, these people had "an unstable state of the nerve system": See Melvin Konner, *The Tangled Wing: Biological Constraints on the Human Spirit,* Henry Holt, 1982.

28 Boas began a deeply entrenched tradition in anthropological research that continues today: See John Tooby and Irven DeVore, "The Reconstruction of Hominid Behavioral Evolution Through Strategic Modeling," in *The Evolution of Human Behavior: Primate Models,* ed. Warren Kinzey, State University of New York, 1987.

28 They and other modern anthropologists embody a research tradition that is based on three fundamental assumptions about the nature of culture: See Donald E. Brown, *Human Universals,* McGraw-Hill, 1991.

30 A *tabula rasa* model of the human mind is a "totalitarian's dream": Noam Chomsky, quoted in Donald Symons, *The Evolution of Human Sexuality,* Oxford University Press, 1979.

31 The ability to learn is also an innate, evolved trait, which takes place within the biological confines of the brain: See Leda Cosmides and John Tooby, "Evolutionary Psychology and the Generation of Culture, Part II: A Computational Theory of Social Exchange," *Ethology and Sociobiology,* vol. 10, 1989.

31 A computer that is purchased for the purpose of processing payroll checks: See S. Pinker and P. Bloom, "Natural Language and Natural Selection," *Behavioral and Brain Sciences,* vol. 13, 1990.

32 "There is no such thing as a 'general problem' ": See Donald Symons, "If We're All Darwinians, What's the Fuss About?" in *Sociobiology and Psychology: Ideas, Issues, and Applications,* ed. Charles Crawford et al., Lawrence Erlbaum, 1987.

33 It is not the moment for the brain to be sifting through the pros and cons of a huge number of possible responses: See John Dupre, ed., *The Latest on the Best: Essays on Evolution and Optimality,* MIT Press, 1987.

33 What kind of psychological mechanisms evolved in the brain that resulted in our ancestors' making the right choices?: See Donald Symons, "If We're All Darwinians, What's the Fuss About?" in *Sociobiology and Psychology: Ideas, Issues, and Applications,* ed. Charles Crawford et al., Lawrence Erlbaum, 1987.

34 These mechanisms regularly produced beneficial behavior during the evolution of our ancient ancestors—and we inherited them: See Donald Symons, *The Evolution of Human Sexuality,* Oxford, 1979.

44 Men and women have slightly different evolved mental specializations for how they deal spatially with their environment: See Irwin Silverman and Marion Eals, "Sex Differences in Spatial Abilities: Evolutionary Theory and Data," in *The Adapted Mind,* ed. Jerome Barkow, Leda Cosmides, and John Tooby, Oxford University Press, 1992.

45 In another study, men and women learned to maneuver a spot of light through a complex maze that appeared on a computer screen and to navigate through a life-size maze: This research was done by Thomas Bever of the University of Rochester.

THE SOCIAL BRAIN

53 " 'What's she going to do about it?' ": D. L. Cheney and R. M. Seyfarth, *How Monkeys See the World: Inside the Mind of Another Species,* University of Chicago Press, 1990.

54 The brains of elephants and whales, for instance, are bigger than those of humans, because these animals' bodies are bigger. See Katharine Milton, "Foraging Behavior and the Evolution of Primate Intelligence" in *Machiavellian Intelligence: Social Expertise and the Evolution of Intellect in Monkeys, Apes, and Humans,* ed. Richard Byrne and Andrew Whiten, Oxford University Press, 1988.

54 The modern human brain is not only bigger than those of most other animals, it is organized differently: "You will not get a human brain if you simply blow up the chimpanzee brain while keeping the same proportions," according to brain researcher Richard Passingham of Oxford University quoted in *New Scientist,* 4 November 1982, p. 288.

56 The first and most obvious was walking upright: In the 1950s, Kenneth Oakley proposed in his book *Man the Tool-Maker* that "when the immediate forerunners of man acquired the ability to walk upright habitually, their hands became free to make and manipulate tools—activities which were in the first place dependent on adequate powers of mental and bodily coordination, but which in turn increase those powers." Quoted in Thomas Wynn, "Tools and the Evolution of Human Intelligence," in *Machiavellian Intelligence: Social Expertise and the Evolution of Intellect in Monkeys, Apes, and Humans,* ed. Richard Byrne and Andrew Whiten, Oxford University Press, 1988.

58 What are primates doing with all those smarts that other animals aren't?:
See Alexander Harcourt, "Alliances in Contests and Social Intelligence,"
in *Machiavellian Intelligence: Social Expertise and the Evolution of Intellect in Monkeys, Apes, and Humans*, ed. Richard Byrne and Andrew
Whiten, Oxford University Press, 1988.

59 Even toddlers will interpret a simple action between two inanimate objects as a social interaction involving goals, desires, and beliefs: See David
Premack, "The Infant's Theory of Self-Propelled Objects," *Cognition*,
vol. 36, 1990.

60 "Vervet monkeys—like characters in a Jane Austen novel—organize their
lives around two principles: to maintain close bonds with kin and to establish good relations with the members of high-ranking families": See
D. L. Cheney and R. M. Seyfarth, *How Monkeys See the World: Inside the
Mind of Another Species*, University of Chicago Press, 1990.

62 "We expect that the monkeys would remain utterly unfazed": See D. L.
Cheney and R. M. Seyfarth, *How Monkeys See the World: Inside the Mind
of Another Species*, University of Chicago Press, 1990.

63 It's a little like people who tell lies only about what they ate for breakfast:
See David Premack, "Does the Chimpanzee Have a Theory of Mind?"
Behavioral and Brain Sciences, vol. 1, pp. 515–26, 1978.

68 The tragic case of autistic children demonstrates that the ability to form
theories about another's mind is an innate part of normal human functioning: See "The Theory of Mind Impairment in Autism: Evidence for
a Modular Mechanism of Development?" in Andrew Whiten, *Natural
Theories of Mind: Evolution, Development, and Simulation of Everyday
Mindreading*, Blackwell, 1991.

69 Being able to build future scenarios of what might happen as a result of
one's actions is a key part of many social exchanges, from deciding whether
to accept an offer for a new job to responding to an insult in a tough bar:
See Richard Alexander, "Evolution of the Human Psyche," in *The Human
Revolution: Behavioral and Biological Perspectives in the Origins of Modern Humans*, ed. Paul Mellars and Chris Stringer, Edinburgh University
Press, 1989.

NICE GUYS FINISH FIRST

78 TIT FOR TAT won again: See Robert Axelrod, *The Evolution of Cooperation*, Basic Books, 1984.

81 "Been watching you a quarter of an hour. Who's winning?": Anecdote from
Getting to Yes, by Roger Fisher and William Ury of the Harvard Negotiation Project, Houghton Mifflin, 1992.

83 The ability to make peace is a crucial component of social relations that
is often overlooked: See Frans de Waal, *Peacemaking Among Primates*,
Harvard University Press, 1989.

86   " 'If it's a ball, throw it right back' ": Ron Luciano, quoted in Robert Axelrod, *The Evolution of Cooperation*, Basic Books, 1984.

WHAT'S GOOD ABOUT FEELING BAD?

94   Flint stopped eating, kept to himself, and sat for hours in the brush, silently rocking back and forth: See Jane Goodall, *Through a Window: My Thirty Years with the Chimpanzees of Gombe*, Houghton Mifflin, 1990.

94   Emotional cues are so important to human survival that a "universal grammar" has evolved in human facial expressions: See Paul Ekman, *Telling Lies: Clues to Deceit in the Marketplace, Politics, and Marriage*, W. W. Norton, 1985.

95   Being known as a blusher might be regarded as a good trait to someone trying to find a trustworthy partner: See Robert Frank, *Passions Within Reason: The Strategic Role of the Emotions*, W. W. Norton, 1988.

101   Children who have chicken pox and are treated with fever-lowering drugs actually take longer to recover from the disease: See "Ancestors May Provide Clinical Answers, Say 'Darwinian' Medical Evolutionists,' *Journal of the American Medical Association*, vol. 269, no. 12, 1993.

103   Even the genes in the fetus itself are in evolutionary conflict: See T. Moore and D. Haig, "Genomic Imprinting in Mammalian Development: A Parental Tug-of-War," *Trends in Genetics*, vol. 7, no. 45, 1991.

104   The biological interests of mother and child result in subtle changes in the mother's psychology as well: See "Pregnancy Sickness as Adaptation: A Deterrent to Maternal Ingestion of Teratogens," by Margie Profet, in *The Adapted Mind*, ed. Jerome Barkow, Leda Cosmides, and John Tooby, Oxford University Press, 1992.

105   Another seemingly modern malady, sadness and depression, may have evolved as a response meant to alleviate the pain of stressful times: See George C. Williams and Randolf M. Nesse, "The Dawn of Darwinian Medicine," *Quarterly Review of Biology*, no. 1, March 1991.

THE EVOLUTION OF LOVE

113   The characteristics judged most important in a mate, whether male or female, are kindness and intelligence: See David M. Buss, "Sex Differences in Human Mate Preferences: Evolutionary Hypotheses Tested in Thirty-seven Cultures," *Behavioral and Brain Sciences*, vol. 12, no. 1, March 1989.

114   Men also have a keen eye for a woman's figure: See Devendra Singh, "Body Shape and Female Attractiveness: The Critical Role of Waist-to-Hip Ratio," *Human Nature*, vol. 4, no. 3, 1993.

116   A woman's desire to mate with an older man appears in a wide variety of cultures: See Douglas Kendrick et al., "Integrating Evolutionary and Social Exchange Perspectives on Relationships: Effects of Gender, Self-Appraisal, and Involvement Level on Mate Selection Criteria," *Journal of Personality and Social Psychology,* vol. 64, no. 6, 1993.

119 Ultimately, whether individuals pursue a short-term or long-term strategy may depend on their perception of what kind of potential mates are in the local environment: See Elizabeth Cashdan, "Attracting Mates: Effects of Paternal Investment on Mate Attraction Strategies," *Ethology and Sociobiology,* 14, 1993.

121 More than 90 percent of people the world over marry at least once—though the Western concept of "lifelong marriage" is by no means the norm in all cultures: See Helen E. Fisher, *The Anatomy of Love: The Natural History of Monogamy, Adultery, and Divorce,* W. W. Norton, 1992.

121 Lucy and her contemporaries lived in social groups like those of modern gorillas: See Richard Klein, *The Human Career: Human Biological and Cultural Origins,* University of Chicago Press, 1989.

122 Only a very small proportion of the sperm ever make it to the egg; the rest are apparently there to serve as "blockers"—also known as "kamikaze sperm"—that slow the progress of other sperm that might be swimming up from behind: See R. Robin Baker and Mark A. Bellis, "Kamikaze Sperm in Mammals?" *Animal Behavior,* vol. 36, no. 3, 1987.

125 Women became potentially dangerous to men's cooperative bonds—an evolutionary legacy that forms the basis for the modern-day "all male" club: See Barbara Smuts, "Male Aggression Against Women: An Evolutionary Perspective," *Human Nature,* vol. 3, no. 1, 1992.

128 New research reveals, however, that in fact, whether—and when—a woman experiences an orgasm has a big effect on the fate of a man's sperm inside her reproductive tract: See R. Robin Baker and Mark A. Bellis, "Human Sperm Competition: Ejaculate Manipulation by Females and a Function for the Female Orgasm," *Animal Behavior,* vol. 45, 1993.

130 The famed "seven-year itch" in modern society may be a vestige of this ancient pattern of serial monogamy among our ancestors: See Helen E. Fisher, *The Anatomy of Love: The Natural History of Monogamy, Adultery, and Divorce,* W. W. Norton, 1992.

130 Control over her reproductive fate was a driving force in producing a woman's evolved psychological mechanisms: See Donald Symons, *The Evolution of Human Sexuality,* Oxford University Press, 1979.

132 The difference in interest in anonymous sex is also reflected in men's and women's sexual fantasies: See Bruce J. Ellis and Donald Symons, "Sex Differences in Sexual Fantasy: An Evolutionary Psychological Approach," *Journal of Sex Research,* vol. 27, no. 4, November 1990.

134 Men are more concerned that their mate has had sex with someone else, while women are typically more concerned with how their mates dispense their resources and attentions: See Margo Wilson and Martin Daly, "The Man Who Mistook His Wife for a Chattel," in *The Adapted Mind,* ed. Jerome Barkow, Leda Cosmides, and John Tooby, Oxford University Press, 1992.

THE BEAST WITHIN

138 The chimpanzees in Gombe had launched another successful attack: See Jane Goodall, *Through a Window: My Thirty Years with the Chimpanzees of Gombe*, Houghton Mifflin, 1990.

140 "Our Descent, then, is the origin of our evil passions!!": Charles Darwin, quoted in John Klama, *Aggression: The Myth of the Beast Within*, John Wiley, 1988.

144 Jealousy is also a prime motivation in wife beating—which claims some 2 million American women as victims each year—and other forms of non-lethal violence: See Barbara Smuts, "Male Aggression Against Women," *Human Nature*, vol. 3, no. 1, 1992.

145 "Homicides by spouses of either sex may be considered slips in this dangerous game": Martin Daly and Margo Wilson, "Homicide," *Science*, 28 Oct. 1988.

147 Studies of child abuse in modern North America reveal that a stepchild is 100 times more likely to be fatally abused by a stepparent: Martin Daly and Margo Wilson, "Homicide," *Science*, 28 Oct. 1988.

GIVING THE MIND A VOICE

161 Just like any other complex, extremely useful part of the human body—such as the eye—language in fact did evolve: See S. Pinker and P. Bloom, "Natural Language and Natural Selection," *Behavioral and Brain Sciences*, vol. 13, 1990.

163 "The lives of our ancestors must have been one long encounter group": See Melvin Konner, *The Tangled Wing: Biological Constraints on the Human Spirit*, Henry Holt, 1982.

170 The mind has been shown to select—and often actively seek out—the kinds of linguistic stimulus it needs to learn language: See Laura A. Petito and Paula F. Marentette, "Babbling in the Manual Mode: Evidence for the Ontogeny of Language," *Science*, 22 March 1991.

170 There are 3,628,800 ways to rearrange the ten words in a sentence such as *Try to rearrange any ordinary sentence consisting of ten words:* Example from Derek Bickerton, *Language and Species*, University of Chicago Press, 1990.

172 Linguists point out that the idea of Eskimos having dozens of different words for snow is a myth: See Geoffrey K. Pullum, *The Great Eskimo Hoax and Other Irreverent Essays on the Study of Language*, University of Chicago Press, 1991.

178 Vervet monkeys have as many as six different alarm calls for the various predators that feed on them: See D. L. Cheney and R. M. Seyfarth, *How Monkeys See the World: Inside the Mind of Another Species*, University of Chicago Press, 1990.

179 A critical difference between vervet calls and human language involves the speaker's and listener's ability to form a theory of mind about each other: See Daniel Dennet, "The Intentional Stance in Theory and Practice," in *Machiavellian Intelligence: Social Expertise and the Evolution of Intellect in Monkeys, Apes, and Humans,* ed. Richard Byrne and Andrew Whiten, Oxford University Press, 1988.

THE CREATIVE EXPLOSION

188 The creative explosion had begun: See Paul Mellars and Chris Stringer, eds., *The Human Revolution: Behavioral and Biological Perspectives in the Origins of Modern Humans,* Edinburgh University Press, 1989.

188 "There was more innovation in the first five minutes of this period than in all the human history leading up to it": See Randall White, "Production Complexity and Standardization in Early Aurignacian Bead and Pendant Manufacture: Evolutionary Implications," in Paul Mellars and Chris Stringer, eds., *The Human Revolution: Behavioral and Biological Perspectives in the Origins of Modern Humans,* Edinburgh University Press, 1989.

193 It is possible for a white person, for instance, to be genetically more similar to an Asian or African than to another white: See Donald E. Brown, *Human Universals,* McGraw-Hill, 1991

197 For a species that thrives on information, long-lived people are an important resource, particularly in ancient societies, which lacked writing: See Randall White, "Thoughts on Social Relationships and Language in Hominid Evolution," *Journal of Social and Personal Relationships,* vol. 2, 1985.

201 Neanderthals were still in Europe thousands of years *after* modern humans arrived on the continent: See Christopher Stringer and Clive Gamble, *In Search of the Neanderthals,* Thames and Hudson, 1993.

204 For the first time, our ancestors had mastered the art of working together to bring down huge amounts of meat on the hoof: See Richard Klein, *The Human Career: Human Biological and Cultural Origins,* University of Chicago Press, 1989.

207 Our *Homo habilis* ancestors were not hunting but *scavenging* for their meat: See Richard Potts, *Early Hominid Activities at Olduvai,* Aldine de Gruyter, 1988.

211 The most magnificent examples of ancient creativity are the majestic paintings our ancestors made on the walls of caves: See Pfeiffer, J. E., *The Creative Explosion: An Inquiry into the Origins of Art and Religion,* Harper and Row, 1982.

THE MODERN EXPERIMENT

225 Societal standards of proper behavior are an important part of helping the group as a whole cooperate: See Robert Axelrod, "An Evolutionary Approach to Norms," *American Political Science Review,* vol. 80, no. 4, December 1986.

229 Why did our ancient ancestors suddenly end their hunting and gathering ways and settle down?: See Joseph Tainter, *The Collapse of Complex Societies*, Cambridge University Press, 1988.

234 The birth of agriculture was not so much a revolution brought about by humans "conquering" plants but a cooperative *coevolution* whereby both plants and humans changed their everyday way of life for their mutual benefit: See David Rindos, *The Origins of Agriculture: An Evolutionary Perspective*, Academic Press, 1984.

237 One big reason we eat beef rather than zebra meat is not because beef is tastier or more nutritious, but simply because cattle will live with people and zebras won't: See Mark Nathan Cohen, *Health and the Rise of Civilization*, Yale University Press, 1989.

241 The Natufians were faced with a choice: Start moving again or find a new way to survive. They turned to agriculture: See O. Bar-Yosef and A. Belfer-Cohen, "From Foraging to Farming in the Mediterranean Levant, in *Transitions to Agriculture in Prehistory*, ed. A. Gebauer and T. Price, University of Wisconsin Press, 1992.

244 Is civilization all that good for us?: See Mark Nathan Cohen, *Health and the Rise of Civilization*, Yale University Press, 1989.

US VERSUS THEM

247 In these scenarios, our ancient ancestors are portrayed as lone heroes: See Misia Landau, "Human Evolution as Narrative," *American Scientist*, vol. 72, May–June 1984.

248 Capping off the whole process, it is implied—a modern human, the icing on the evolutionary cake: See R. A. Foley, ed., *The Origins of Human Behavior*, Unwin Hyman, 1991.

Bibliography

Alexander, Richard D. *The Biology of Moral Systems.* Aldine, 1987.

Axelrod, Robert. *The Evolution of Cooperation.* Basic Books, 1984.

———. "An Evolutionary Approach to Norms." *American Political Science Review,* vol. 80, no. 4, December 1986.

Baker, R. Robin, and Bellis, Mark A. "Human Sperm Competition: Ejaculate Manipulation by Females and a Function for the Female Orgasm." *Animal Behavior,* vol. 45, 1993.

———. "Human Sperm Competition: Ejaculate Adjustment by Males and the Function of Masturbation." *Animal Behavior,* vol. 45, 1993.

———. "Do Females Promote Sperm Competition? Data for Humans." *Animal Behavior,* vol. 40, no. 5, 1990.

———. "Number of Sperm in Human Ejaculates Varies in Accordance with Sperm Competition Theory." *Animal Behavior,* vol. 37, no. 5, 1988.

———. "Kamikaze Sperm in Mammals?" *Animal Behavior,* vol. 36, no. 3, 1987.

Bar-Yosef, O., and Belfer-Cohen, A. "From Foraging to Farming in the Mediterranean Levant." In *Transitions to Agriculture in Prehistory,* edited by A. Gebauer and T. Price. University of Wisconsin Press, 1992.

Barkow, Jerome; Cosmides, Leda; and Tooby, John, eds. *The Adapted Mind.* Oxford University Press, 1992.

Belsky, Jay; Steinberg, Laurence; and Draper, Patricia. "Childhood Experience, Interpersonal Development, and Reproductive Strategy: An Evolutionary Theory of Socialization," *Child Development,* 62, 1991.

Bickerton, Derek. *Language and Species.* University of Chicago Press, 1990.

Brown, Donald E. *Human Universals.* McGraw-Hill, 1991.

Buss, David M. "Sex Differences in Human Mate Preferences: Evolutionary Hypotheses Tested in Thirty-seven Cultures." *Behavioral and Brain Sciences,* vol. 12, no. 1, March 1989.

Buss, David M., and Schmitt, David P. "Sexual Strategies Theory: An Evolutionary Perspective on Human Mating." *Psychological Review,* vol. 100, no. 2, 1993.

Byrne, Richard, and Whiten, Andrew, eds. *Machiavellian Intelligence: Social Expertise and the Evolution of Intellect in Monkeys, Apes, and Humans.* Oxford University Press, 1988.

Cartmill, Matt. "Four Legs Good, Two Legs Bad." *Natural History*, vol. 11, 1983.

Cashdan, Elizabeth. "Attracting Mates: Effects of Paternal Investment on Mate Attraction Strategies." *Ethology and Sociobiology*, vol. 14, 1993.

Chagnon, Napoleon A. *Yanomamo: The Last Days of Eden.* Harcourt Brace Jovanovich, 1992.

Cheney, D. L., and Seyfarth, R. M. *How Monkeys See the World: Inside the Mind of Another Species.* University of Chicago Press, 1990.

Cohen, Mark Nathan. *Health and the Rise of Civilization.* Yale University Press, 1989.

Cosmides, Leda. "The Logic of Social Exchange: Has Natural Selection Shaped How Humans Reason?" *Cognition*, vol. 31, 1989.

Cosmides, Leda, and Tooby, John. "Evolutionary Psychology and the Generation of Culture, Part II: A Computational Theory of Social Exchange." *Ethology and Sociobiology*, vol. 10, 1989.

Cronin, Helena. *The Ant and the Peacock.* Cambridge University Press, 1991.

Daly, Martin, and Wilson, Margo. "Killing the Competition: Female/Female and Male/Male Homicide," in *Human Nature*, vol. 1, no. 1, 1990

————. "Evolutionary Social Psychology and Family Homicide." *Science*, vol. 242, October 28, 1988.

————. *Homicide.* Aldine de Gruyter, 1988.

————. *Sex, Evolution, and Behavior.* Wadsworth, 1983.

Degler, Carl N. *In Search of Human Nature: The Decline and Revival of Darwinism in American Social Thought.* Oxford University Press, 1991.

Deutsch, Diana. *The Psychology of Music.* Academic Press, 1982.

Dupre, John, ed. *The Latest on the Best: Essays on Evolution and Optimality.* MIT Press, 1987.

Ekman, Paul. *Telling Lies: Clues to Deceit in the Marketplace, Politics, and Marriage.* W. W. Norton, 1985.

Ellis, Bruce J., and Symons, Donald. "Sex Differences in Sexual Fantasy: An Evolutionary Psychological Approach." *Journal of Sex Research*, vol. 27, no. 4, November 1990.

Falk, Dean. *Braindance: New Discoveries About Human Origins and Brain Evolution.* Henry Holt, 1992.

Fisher, Helen E. *The Anatomy of Love: The Natural History of Monogamy, Adultery, and Divorce.* W. W. Norton, 1992.

Foley, R. A., ed. *The Origins of Human Behavior.* Unwin Hyman, 1991.

Frank, Robert. *Passions Within Reason: The Strategic Role of the Emotions.* W. W. Norton, 1988.

Gazzaniga, Michael S. *Nature's Mind: The Biological Roots of Thinking, Emotions, Sexuality, Language, and Intelligence.* Basic Books, 1992.

Goodall, Jane. *Through a Window: My Thirty Years with the Chimpanzees of Gombe.* Houghton Mifflin, 1990.

Harcourt, Alexander, and de Waal, Franz. *Coalitions and Alliances in Humans and other Animals.* Oxford University Press, 1992.

Hass, Jonathan. *The Evolution of the Prehistoric State,* Columbia University Press, 1982.

Heltne, Paul G., and Marquardt, Linda A. *Understanding Chimpanzees.* Harvard University Press, 1989.

Humphrey, Nicholas. *Consciousness Regained: Chapters in the Development of Mind.* Oxford University Press, 1984.

Hinde, Robert A., and Groebel, Jo. *Cooperation and Prosocial Behavior.* Cambridge University Press, 1991.

Kendrick, Douglas, et al. "Integrating Evolutionary and Social Exchange Perspectives on Relationships: Effects of Gender, Self-Appraisal, and Involvement Level on Mate Selection Criteria." *Journal of Personality and Social Psychology,* vol. 64, no. 6, 1993.

Kitcher, Philip. *Vaulting Ambition: Sociobiology and the Quest for Human Nature.* MIT Press, 1987.

Klama, John. *Aggression: The Myth of the Beast Within.* John Wiley, 1988.

Klein, Richard. *The Human Career: Human Biological and Cultural Origins.* University of Chicago Press, 1989.

Konner, Melvin. *The Tangled Wing: Biological Constraints on the Human Spirit.* Henry Holt, 1982.

Landau, Misia. "Human Evolution as Narrative." *American Scientist,* vol. 72, May–June 1984.

Lieberman, D. E. "Seasonality and Gazelle Hunting at Hayonim Cave: New Evidence for Sedentism During the Natufian." *Paleorient,* vol. 17, no. 1, 1991.

Loy, James D., and Peters, Calvin B. *Understanding Behavior: What Primate Studies Tell Us About Human Behavior.* Oxford University Press, 1991.

Manson, Joseph H., and Wrangham, Richard W. "Intergroup Aggression in Chimpanzees and Humans." *Current Anthropology,* vol. 32, no. 4, 1991.

Mellars, Paul, and Stringer, Chris, eds. *The Human Revolution: Behavioral and Biological Perspectives in the Origins of Modern Humans.* Edinburgh University Press, 1989.

Moore, T., and Haig, D. "Genomic Imprinting in Mammalian Development: A Parental Tug-of-War." *Trends in Genetics,* vol. 7, no. 45, 1991.

Nesse, Randolf M. "Evolutionary Explanation of the Emotions." *Human Nature,* vol. 1, no. 3, 1990.

Parker, Sue Taylor, and Gibson, Kathleen Rita. *"Language" and Intelligence in Monkeys and Apes.* Cambridge University Press, 1990.

Pines, David. *Emerging Syntheses in Science.* Addison-Wesley, 1988.

Petito, Laura A. "On the Autonomy of Language and Gesture: Evidence from the Acquisition of Personal Pronouns in American Sign Language." *Cognition,* vol. 27, no. 1, October 1987.

Petito, Laura A., and Marentette, Paula F. "Babbling in the Manual Mode: Evidence for the Ontogeny of Language." *Science,* 22 March 1991.

Pfeiffer, J. E. *The Creative Explosion: An Inquiry into the Origins of Art and Religion.* Harper and Row, 1982.

Pinker, S., and Bloom, P. "Natural Language and Natural Selection." *Behavioral and Brain Sciences,* vol. 13, 1990.

Potts, Richard. *Early Hominid Activities at Olduvai.* Aldine de Gruyter, 1988.

Potts, Richard, and Shipman, P., "Cutmarks Made by Stone Tools on Bones from Olduvai Gorge, Tanzania." *Nature,* no. 291, 1981.

Premack, David. "The Infant's Theory of Self-Propelled Objects." *Cognition,* vol. 36, 1990.

———. "Does the Chimpanzee Have a Theory of Mind?" *Behavioral and Brain Sciences,* vol. 1, pp. 515–26, 1978.

Pullum, Geoffrey K. *The Great Eskimo Hoax and Other Irreverent Essays on the Study of Language.* University of Chicago Press, 1991.

Richards, Robert. *Darwin and the Emergence of Evolutionary Theories of Mind and Behavior.* University of Chicago Press, 1987

Rindos, David. *The Origins of Agriculture: An Evolutionary Perspective.* Academic Press, 1984.

Robinson, Michael H., and Tiger, Lionel. *Man and Beast Revisited.* Smithsonian Institution Press, 1991.

Singh, Devendra. "Body Shape and Female Attractiveness: The Critical Role of Waist-to-Hip Ratio." *Human Nature,* vol. 4, no. 3, 1993.

Smith, Eric A., and Winterhalder, Bruce. *Evolutionary Ecology and Human Behavior.* Aldine de Gruyter, 1992.

Smith, John Maynard. *Evolution and the Theory of Games.* Cambridge University Press, 1982.

Smuts, Barbara. "Male Aggression Against Women: An Evolutionary Perspective." *Human Nature,* vol. 3, no. 1, 1992.

Smuts, Barbara B., and Smuts, Robert W. "Male Aggression and Sexual Coercion of Females in Nonhuman Primates and Other Mammals: Evidence and Theoretical Implications." In *Advances in the Study of Behavior,* vol. 22, edited by Peter J. Slater. Academic Press, 1992.

Symons, Donald. "Adaptiveness and Adaptation," *Ethology and Sociobiology,* vol. 11, no. 4/5, 1990.

———. "If We're All Darwinians, What's the Fuss About?" In *Sociobiology and Psychology: Ideas, Issues, and Applications,* edited by Charles Crawford et al. Lawrence Erlbaum, 1987.

———. *The Evolution of Human Sexuality.* Oxford, 1979.

Stringer, Christopher, and Gamble, Clive. *In Search of the Neanderthals.* Thames and Hudson, 1993.

Tainter, Joseph. *The Collapse of Complex Societies.* Cambridge University Press, 1988.

Thornhill, Randy, and Gangestad, Steven W. "Human Facial Beauty: Averageness, Symmetry, and Parasite Resistance." *Human Nature,* vol. 4, no. 3, 1993.

Tiger, Lionel. *The Pursuit of Pleasure.* Little Brown, 1992.

Tooby, John, and Cosmides, Leda. "Evolutionary Psychology and the Generation of Culture, Part 1: Theoretical Considerations." *Ethology and Sociobiology,* vol. 10, 1989.

————. "On the Universality of Human Nature and the Uniqueness of the Individual: The Role of Genetics and Adaptation." *Journal of Personality,* vol. 58, 1990.

————. "The Past Explains the Present: Emotional Adaptations and the Structure of Ancestral Environments," in *Ethology and Sociobiology,* vol. 11, no. 4/5, 1990.

Tooby, John, and DeVore, Irven. "The Reconstruction of Hominid Behavioral Evolution Through Strategic Modeling." In *The Evolution of Human Behavior: Primate Models,* edited by Warren Kinzey. State University of New York, 1987.

Trinkhaus, Erik, and Tompkins, Robert L. "The Neanderthal Life Cycles: The Possibility, Probability, and Perceptibility of Contrasts with Recent Humans." In *Primate Life History and Evolution,* edited by C. Jean de Rousseau. Wiley-Liss, 1990.

de Waal, Frans. *Peacemaking Among Primates.* Harvard University Press, 1989.

White, Randall. "Thoughts on Social Relationships and Language in Hominid Evolution." *Journal of Social and Personal Relationships,* vol. 2, 1985.

————. "Production Complexity and Standardization in Early Aurignacian Bead and Pendant Manufacture: Evolutionary Implications." In *The Human Revolution: Behavioral and Biological Perspectives in the Origins of Modern Humans,* edited by Paul Mellars and Chris Stringer. Edinburgh University Press, 1989.

————. "Visual Thinking in the Ice Age." *Scientific American,* July 1989.

Whiten, Andrew. *Natural Theories of Mind: Evolution, Development, and Simulation of Everyday Mindreading.* Blackwell.

Williams, George C., and Nesse, Randolf M. "The Dawn of Darwinian Medicine." *Quarterly Review of Biology,* no. 1, March 1991.

Wilson, Margo, and Daly, Martin. "The Man Who Mistook His Wife for a Chattel." In *The Adapted Mind,* edited by Jerome Barkow, Leda Cosmides, and John Tooby. Oxford University Press, 1992.

INDEX

About the Author

WILLIAM F. ALLMAN is a senior writer at *U.S. News & World Report*, where he covers anthropology, brain science, and human behavior. His previous books are *Apprentices of Wonder: Inside the Neural Network Revolution*, which chronicles the recent development of intelligent, brain-style computers, and *Newton at the Bat: The Science in Sports*. His articles have appeared in *Esquire, The Washington Post, OMNI, Technology Review*, and other magazines. He lives in Washington, D.C.